Geographic Information Systems

GEOGRAPHIC INFORMATION SYSTEMS: DEVELOPMENTS AND APPLICATIONS

Edited By Les Worrall

Belhaven Press
(a division of Pinter Publishers)
London **and New** York

© Les Worrall and Contributors, 1990

First published in Great Britain in 1990 by Belhaven Press (a division of Pinter Publishers), 25 Floral Street, London WC2E 9DS

British Library Cataloguing in Publication Data
A CIP catalogue record for this book is available from the British Library

ISBN 1 85293 140 X

For enquiries in North America please contact
PO Box 197, New York, NY 10533

Library of Congress Cataloging-in-Publication Data

Geographic information systems : developments & applications / edited
  by Les Worrall
      p.   cm.
   Includes index.
   ISBN 1-85293-140-X
   1. Geographic information systems.   I. Worrall, Les.
  G70.2.G45   1990
  910′.285--dc20   90-44434
                                                                       CIP

Typeset by Communitype Communications Ltd.
Printed and bound in Great Britain by Biddles Ltd of Guildford and Kings Lynn

CONTENTS

LIST OF FIGURES

List of Tables

List of Contributers

Dr Mark Birkin, School of Geography, University of Leeds, Leeds, LS2 9JT, UK.

Dr Graham Clarke, School of Geography, University of Leeds, Leeds, LS2 9JT, UK.

Dr Martin Clarke, School of Geography, University of Leeds, Leeds, LS2 9JT, UK.

Prof. Barry J. Garner, School of Geography, University of New South Wales, P.O. Box 1, Kensington, NSW2033, Australia.

Iain Gault, ICL UK Ltd, 127 Hagley Road, Birmingham, B16 8LD, UK.

Dr S.C.M. Geertman, Faculty of Geographical Sciences, University of Utrecht, P.O. Box. 80.115, 3508TC, Utrecht, the Netherlands.

Dr Karl E. Kim, Department of Urban and Regional Planning, University of Hawaii at Manoa, 2424 Maile Way, Honolulu, Hawaii 96822, USA.

Dr Thomas L. Millette, Department of Geography, University of Maryland, Social Science Building, Baltimore, Maryland, 21228, USA.

Dr Larry G. O'Brien, Special Needs Information Research Unit, Department of Geography, University of Newcastle-upon-Tyne, Newcastle, NE1 7RU, UK.

David Peutherer, Chief Executive's Department, Strathclyde Regional Council, India Street, Glasgow, G2 4PF, UK.

Prof. Dr Henk Scholten, National Institute for Public Health and Environmental Protection, Antonie van Leeuwenhoeklaan 9, P.O. Box. 1, 3720 BA, Bilthoven, The Netherlands and Department of Informatics, Free University of Amsterdam, Amsterdam, The Netherlands.

Dr F.J. Toppen, Faculty of Geographical Sciences, University of Utrecht, P.O. Box. 80.115, 3508TC, Utrecht, The Netherlands.

Dr Maurits van der Vlugt, Faculty of Geographical Sciences, University of Utrecht, P.O. Box. 80.115, 3508TC, Utrecht, The Netherlands.

Prof. Alan Wilson, School of Geography, University of Leeds, Leeds, LS2 9JT, UK.

Dr Les Worrall, Department of Civic Design, P.O. Box. 147, University of Liverpool, Liverpool, L69 3BX, UK and Policy Unit, Wrekin Council, P.O. Box. 213, Malinslee House, Telford, TF3 4LD, UK.

Dr Anthony Gar-On Yeh, Centre for Urban Studies and Urban Planning, University of Hong Kong, Pokfulam Road, Hong Kong.

Chapter One
GIS: prospects and challenges
Les Worrall

Introduction

In the last five years, there has been an accelerating growth in the volume, variety, power and sophistication of the computer-based tools available to support urban and regional planning and policy-making: recent progress in geographic information systems (GIS) is at the forefront of these developments. As with any technological innovation, the key element in the success of GIS will be the extent to which the array of techniques and technology are practically applied: with GIS, it will be their successfulness in making social decision-making processes better, stronger and more accessible which will determine their longevity. In the early phases of any technological innovation, the prime concern tends to be focused on the production of techniques rather than with their consumption: GIS is no exception to this general rule. Consequently, the fundamental objective of this book is to focus on the issues which surround the transfer of GIS technology and techniques into those organisations concerned with the development, analysis and implementation of public policy.

While GIS is a tool of considerable promise and potential to public policy-making, some authors have raised doubts about its ultimate applicability in policy-making environments (see Worrall, 1990d). Couclelis (1989) has argued that GIS is 'not yet fully attuned to urban and regional planning needs' and that GIS has had more of an impact on 'lower-order' planning activities (such as engineering and spatial design) than on 'higher order' activities (such as strategic planning and policy-making). The development of GIS has been criticised as being 'technology-led' rather than 'planning-led' (Batty, 1988). Breheney (1987) has raised a number of fundamental issues about the role of analytical techniques and methods in the strategic planning process and, in particular, he has criticised much academic research as being motivated more by theoretical and technical issues than by policy issues — this criticism is especially relevant to the current stage in the development of GIS.

Worrall (1990a,b,c) has also raised questions about the quality and usefulness of the information systems available to support GIS development. Birkin (Birkin *et al.*, 1987) has argued that analytical modelling techniques have been inadequately incorporated into GIS, while Newkirk (1987) and Klosterman (1989) have both remarked that many of the applications of computer-based analysis in practical planning environments have been unsophisticated to the point of being rudimentary. Smith (1989) has noted that the problems inhibiting the application of GIS are more often institutional and managerial rather than technical: this is a concern that few GIS protagonists have considered. Finally, French and Wiggins (1989) have identified that the shortage of information specialists in planning agencies in the USA may act as an impediment to the wider adoption of GIS techniques.

It is apparent that the issues surrounding the transfer of GIS technology and techniques into public policy-making environments are complex and highly interacting: these issues include concerns over the inherent suitability of GIS to policy-making processes, concerns about information quality and usability and concerns about the ability of policy-making organisations to exploit the techniques fully due both to the fact that the development of GIS may cause shifts in the power structures within organisations and to the lack of trained personnel. The role of this book is to explore these issues.

Themes and contents

The chapters contained in this book are intended to pursue the issues which surround the application of GIS in practical policy-making environments. The emphasis throughout is not on describing GIS technology itself but on examining the issues which surround its use and on reviewing the contribution it has made to urban and regional planning and policy-making. Given that the nature of the planning system into which GIS is to be implanted will have a major effect on the success with which GIS can be applied, the examples used throughout the book are drawn from several different countries each of which has a differing approach to spatial planning and policy-making. The examples used include Hong Kong where the government has a high degree of control over the urban development process, the Netherlands where there is a well established physical planning regime, the UK where the tension between central and local government is creating a turbulent environment for the adoption of GIS, Australia where LIS are well developed and being used as platforms for the development of state and city-wide GIS and the USA where GIS techniques are perhaps the best developed. Given the concern with both the integration of model-based analysis within GIS and with the development of appropriate spatial information systems, two chapters based on research at the Leeds University, School of Geography and on the National Online Manpower Information System (NOMIS) at Durham University have been included to address these issues with reference to practical examples.

The chapter by Scholten and van der Vlugt contains a review of GIS applications throughout Europe. Generally, their review reinforces the concern that most GIS have been applied to 'lower order' planning activities in which the domain of interest has been well defined: the applications they review are generally task-orientated rather than process-orientated. Nevertheless, the range of GIS applications examined is extensive and serves to demonstrate that GIS is a powerful spatial planning tool applicable to a wide range of areas: these include forestry management, environmental monitoring, site evaluation, marketing, health care planning, spatial epidemiology, emergency planning, risk assessment and transportation planning. In addition to their review of applications, Scholten and van der Vlugt develop a useful categorisation of GIS users, each of which has differing needs which must be carefully considered in the development and application of GIS. Their typology serves to show that the ability to interrogate spatial information will not be equally distributed throughout any organisation and this raises the question of how GIS can be made more accessible to a variety of users: this is a topic which is also addressed in the chapters by Geertman and Toppen and by Gault and Peutherer.

Before discussing the future for GIS as they see it, Scholten and van der Vlugt identify some problems with the current stage in the development of GIS. They are concerned that GIS specialists all too often see GIS development as an end in itself and are often tempted to apply their resources to answering irrelevant or low priority questions. More important, they identify that GIS endeavour will only become more focused on questions of high social priority when decision-makers (both political and senior executive) are encouraged to ask the right questions. Scholten and van der Vlugt also see GIS as providing a unifying framework for the more integrated development of social geoscience: indeed, some of the examples they cite (such as Openshaw's (*et al.*, 1987) work on the identification of leukaemia clusters) would seem to support this view. At the end of their review, two GIS development streams are identified: the first stream is directly relevant to improving the applicability of GIS and is concerned with the development of spatial analysis techniques (this is pursued in the chapter by Birkin and his colleagues from Leeds University) while the second stream (which includes topics such as the incorporation of artificial intelligence techniques and 'fuzzy logic' into GIS) is more orientated towards pure technical issues.

In a second review chapter/Garner examines the development and application of GIS in urban and regional planning in Australia. He shows that major steps have been taken in Australia in the development of land information systems (LIS), which despite the fact that they have a more limited analytical potential than GIS, are providing the necessary platform for the development of GIS for both state and city-level planning authorities in Australia. This development path will include the incorporation of more complex analytical procedures (particularly model-based analysis), the use of a wider range of linked data sets and a move away from natural resource and environmental management applications towards an increased role in social and economic planning. The LIS applications which Garner describes are currently geared primarily to land management: he shows that the more systematic collection and maintenance of land information has led to a reduction in the multiple holding of often inaccurate data and that the sharing of data has led to the more effective management of information both within and between organisations. Garner shows that national-level institutional arrangements are necessary for the more coordinated development of LIS and GIS: in Australia, the creation of the Australian Land Information Council in 1986 provided such a legislatively ordained national framework for the development of state-wide LIS/

While Garner demonstrates that the application of GIS is somewhat limited in local government in Australia, his discussion of GIS-linked transportation modelling in the Commonwealth Scientific and Industrial Research Organisation (CSIRO) shows that this field of research is well developed. The 'Transport Planning Microscope' (TPM) developed by Taylor (1989) is described as an interactive PC-based system which contains GIS software linked to a connected hierarchy of transportation modelling packages which offers an array of new opportunities for integrated data analysis and for measuring the performance of systems (the use of system performance indicators is also addressed in the chapter by Birkin *et al.*). Several authors (Scholten and van der Vlugt in this volume, Cook *et al.*, 1989) have commented upon the fact that transportation planning is a field in which GIS and model-based analysis have been successfully linked: Garner reinforces the view that transportation planning and modelling is a field from which other GIS application areas may have much to learn.

In the next two chapters, Millette and Kim describe the developments that have taken place in the application of GIS in the states of Vermont and Hawaii respectively. According to Millette, the Vermont GIS is the first state-wide GIS to be legislatively mandated and funded in the USA. Given the scale of the task of developing the necessary data bases and the planned five year programme for assembling the various data layers, a long term state commitment to funding is seen as essential for the successful development of the system. In addition to the legislative mandate and funding programme (the finance for which is to be obtained by the levying of a real estate transfer tax), Millette shows that the Vermont GIS has been built on the basis of a strong commitment from the various state-wide, regional and local planning agencies. This commitment grew out of a study sponsored by the state governor which examined the issues surrounding the future economic and physical development of the state on the one hand against the need to conserve the local environment on the other.

Millette shows that because the Vermont GIS was intended to support planning activities at the state, regional and local levels, it was necessary to develop a means of delivering this distributed form of planning support to all the agencies involved. After contrasting the advantages of centralised and distributed approaches to providing GIS support, Millette describes how it was decided to use Vermont's twelve regional planning commissions (RPCs) to deliver GIS support. Millette argues that the RPC model, with support from the University of Vermont, was favoured over a centralised state-wide approach for two main reasons: first, that the RPCs already provided 'traditional' planning support to local planning agencies and, second, that a distributed approach would facilitate better public access to a system which was more locally focused. However, in order to ensure the state-wide integrity of the system and the adherence to common standards, an Office of Geographic Information Services has been established. Clearly, the message to be learned from Millette's description of the Vermont case study is that GIS will make a more effective contribution to planning when, because of a strong political commitment to the development of GIS and the belief in its potential benefits, GIS becomes an integral part of administration and is adequately funded over the long term.

The development and application of GIS in Hawaii, as described by Kim, is examined from three governmental levels (state, county and municipality). At the state level, Kim points out that a commitment was made in 1987 to develop a GIS which would permit various public and private agencies access to data, give them the ability to extract the information they need and allow them to analyse information 'according to individual requirements'. Kim describes in some detail the mapping base of the system (the USGS 1:24,000 base maps — which have emerged as the standard for planning purposes in Hawaii and elsewhere in the USA). He also describes the developments and enhancements that are proposed for these maps by both the USGS and by the data collection activities of the State of Hawaii. Kim shows that, in terms of prime usage, the state level GIS is orientated mainly to resource management given the massive development pressures that Hawaii has faced over the last twenty years.

At the municipal and county level (using the example of the city and county of Honolulu), the uses to which GIS have been applied are different from the prevailing state-level uses. The city GIS is more orientated towards permit management, administrative decision-making and service planning. Kim remarks that an important aspect in the development of the Honolulu GIS was that it was procured as a 'turnkey' solution in which the Environmental Systems Research Institute (ESRI) was contracted to design a GIS, to convert various maps and data bases, to develop applications and to deliver a fully operational GIS to the city authorities. Kim argues that the future of GIS and its success in practical decision-making environments is inextricably linked to concerns about the quality of data, access to data, data maintenance, training and education and the development of better GIS-centred planning methods. However, these are issues which Hawaii shares with many other locations.

The chapters by Millette and Kim show that GIS applications are well developed in the USA and that GIS have become, or are increasingly becoming, a routine part of both administration and planning in the USA. The chapter by Gault and Peutherer shows that in the UK not as much progress has been made and that this is due to the sometimes turbulent relationship between central and local government in the UK, to inadequacies in the data sets available and often to a lack of commitment at the highest decision-making levels within local government. Nevertheless, by focusing on these organisational issues, Gault and Peutherer in their reviews of GIS development in Birmingham City Council and Strathclyde Regional Council, show that some progress has been made. In the Birmingham case, a structured pilot exercise was devised to test a range of application types relevant to different levels in the UK local government management hierarchy (operational, managerial and strategic). They demonstrate that the pilot exercise was structured both by the type of spatial data being used (i.e. point, network and polygon) and by the means of data capture or transfer (i.e. mainframe to PC, PC to PC, manual collection and digitising to assess the relative costs of data base creation).

In the Strathclyde case study, Peutherer shows that considerable emphasis has been placed on the development of strategic information systems to support the policy development activities of the Regional Council. The strategic information base has been developed by making progress on three fronts simultaneously: this has been achieved by designing local, primary data collection systems, by capturing data from external organisations and by spatially referencing internal, operational data bases. In the Strathclyde case study, unlike many others, the decision to obtain GIS software followed, rather than led, the development of strategic information systems. Gault and Peutherer reiterate many of the points made by Scholten and van der Vlugt in their discussion of the differing needs of the various GIS user types and they conclude by identifying the need for a clear organisation-wide information strategy and the need for GIS specialists systematically to build the commitment of political and senior executive decision-makers.

In a second case study based on the UK, the emphasis moves away from a direct applications environment to examine the development of model-based GIS, based on the work of Birkin and his colleagues in the School of Geography at the University of Leeds. In the chapters by Scholten and van der Vlugt and by Garner, it is concluded that a logical and essential development of GIS is their successful integration with model-based analysis. In the chapter by Birkin et al. the situation is reversed. The School of Geography at the University of Leeds has a long history of innovation in model-based analysis and, in their case, GIS techniques have been seen as a means of improving the presentational aspects of their existing work. Birkin and colleagues argue that the Leeds approach is distinctive in two respects: first, it is model-based rather than being data-based, as are many of the systems which have been developed recently, and, second, the Leeds approach uses a variety of software products which, it is argued, makes the systems which have been developed more suited to the needs of particular users and more powerful. The authors argue that their approach is more flexible and more responsive to user needs than solutions derived from 'off-the-shelf' GIS packages.

In addition to the straightforward modelling of systems, the Leeds geographers are committed to the idea of building system performance indicators into their models: this concept is developed further in their discussion of the West Yorkshire IGIS ('Intelligent' GIS) and in the application of their techniques to labour market analysis and retail planning. The models developed by the Leeds School of Geography are highly information intensive, and given the fact that many of the information items needed to support their models do not exist at all or exist only as national or regional aggregates, Birkin and his colleagues have developed powerful techniques to simulate the data they require. While GIS can only be considered an adjunct to the research presented in this chapter, the simulation techniques, model-based analytical techniques and the development of performance indicators within models are all shown to be powerful devices which merit wider application.

In the next two chapters, attention is focused on the use of GIS in monitoring the supply of land for urban development in Hong Kong and allocating land for housing development in the Netherlands. Yeh paints a picture of the need to plan to accommodate considerable development pressure in his case study of Hong Kong: Hong Kong has an average population density of over 5,000 people per square kilometre (though this approaches 140,000 people per square kilometre in one residential neighbourhood): it is one of the most densely populated city-states in the world. Between 1931 and the present day, the population increased from 300,000 to almost 6,000,000: this volume and rate of population growth has exerted considerable pressure on housing, on transportation systems and on community facilities. As a result of this considerable pressure on land, and the consequent need to manage land resources effectively, Yeh describes how the Hong Kong government developed the computer-based systems which could store, retrieve and analyse land information and provide a direct input to urban planning and urban management in Hong Kong. As with the establishment of the ALIC in Australia, and as a direct result of government legislation, a Land Information Centre (LIC) was established in Hong Kong in 1987 and a distributed network of workstations was established throughout various government agencies and district planning offices in early 1989.

Yeh describes the structure and the technological aspects of the land information system in Hong Kong before going on to describe how the system is used in conjunction with population and household forecasts to calculate quite mechanistically (though quite accurately given the nature of the housing position in Hong Kong) the number of housing units needed in a given period of time. The estimation of housing needs and the provision of housing is made much easier in Hong Kong by the fact that the government is the major landowner and that most of the newly developed land results almost exclusively from the government reclamation of land. Despite the relative simplicity of the modelling techniques used, Yeh shows that the Hong Kong LIS has made a significant contribution to urban planning and urban management in Hong Kong.

While Hong Kong is one of the most densely populated cities in the world, the Randstad Holland (which is a polynuclear urban area comprising Amsterdam, The Hague, Rotterdam and several other major cities in the Netherlands) is the most densely populated part of the Netherlands. The urban morphology of this area makes planning in the Randstad Holland very difficult particularly given increasing population and development pressures on the one hand and the need to maintain open space standards and preserve the living environment on the other. It is in this complex planning arena that a GIS-based approach to allocating land for the development of over 1,000,000 housing units between the years 1990 and 2015 has been developed by Geertman and Toppen. The technique they have developed in their pilot study is based on the use of various overlays derived from several nationally available data sources. In later versions of their approach, they propose to extend the number of planning criteria used (that is to increase both the number of overlays and planning conditions) in order to make their prognoses more sensitive to national and municipal planning goals. One particular problem that the authors identified was the need to make GIS more user-friendly. This has been achieved in the Netherlands by the development of a 'Spatial Analysis Tool' or SPAT (see Zevenbergen, 1989) which allows the user (in this case predominantly regional planners) to select, to classify, to query, to plot map layers and to perform basic analytical functions without having to learn more complex GIS command languages. Geertman and Toppen suggest that the further development of tools such as SPAT are an effective means of broadening the user base of GIS.

In the last chapter, attention is turned to some of the data and information issues which underpin the use of GIS in urban and regional planning and policy-making. O'Brien uses NOMIS (National Online Manpower Information System), which has been described as the largest GIS in the UK, as a basis for examining a range of issues. He demonstrates how NOMIS has been used as a basis for policy formulation, programme delivery and resource targeting before discussing some of the academic and applied research that the system has also supported. He uses Nijkamp and de Jong's (1987) concept of 'orgware' to describe NOMIS as a system which has been consciously designed to fit into existing organisational structures and procedures: he states that many computer systems have failed simply because their degree of fit within their receptor organisations has been poor.

O'Brien concludes by examining some of the problems inherent in the use of GIS and identifies a number of problems that the unwary GIS user may encounter because of his unfamiliarity with the nature and the definitional assumptions used in the data sets he is using. These potential pitfalls include problems which arise because of the choice of basic spatial unit (see Visvalingham, 1988a,b) and also those problems which arise because many of the definitions used in administrative information systems, censuses and surveys are often ill-considered, variable through time and variable in social meaning: O'Brien illustrates this point by reference to the changing definition of the term 'disability' and reminds us all that such terms, which are often used unthinkingly in all forms of spatial analysis, derive their relevance from the effect they have on peoples' lives.

Directions

The advance of GIS will lie in its further application in practical policy-making environments: the case studies by Kim on Hawaii and Millette on Vermont show that both a high level of political commitment and assured funding for the acquisition of hardware and software and for data base creation are essential if GIS is to become an integral part of decision-making, planning and administration. One of the main points to emerge from this book is the need to overcome the problem of fragmented development in analytical model-building, information system development, technological development and applied policy analysis by taking a more integrated approach based on a clearer statement of the needs of decision-makers. The whole process must be policy-driven, and there is a need to be aware of changing approaches to urban and regional management which are now being developed. Current approaches to urban management tend to be process-orientated, concerned more with creating a form of government which is more open and democratic and in which the public and private sector can jointly develop and implement programmes. By and large, the concern is not with the derivation and selection of optimal strategies — it is more orientated towards a process of negotiation and bargaining and the evaluation of competing solutions. This has major implications for the devices needed to support distributed decision-making and it is here that GIS can make a fundamental contribution. In addition, GIS is not solely about doing old things quicker and producing more picturesque output, it is about identifying new solutions to problems and it must contain the realisation that all social science perspectives are just as relevant as each other and just as valid if they can contribute to the solution of social problems.

GIS will also only achieve its full potential if there is a far greater emphasis on developing strategic information systems: urban management involves many kinds of data and many kinds of processing — hence hardware and software must be flexible and adaptable to a variety and hierarchy of needs. The needs of policy-makers must be used as a contextual framework for the development of urban and regional data models and for the development of information strategies which integrate numerical, spatial, textual and graphic information. The development of such an approach requires the commitment of senior executives: yet, based on past failures, it is very often these senior executives who are most sceptical about the espoused benefits of new technology. Their commitment will only be obtained if there is a change in emphasis away from a technologically-driven production-orientation towards a GIS development path which is sensitive to the context in which policy decisions are made and geared to meeting the needs of strategic decision-makers.

Bibliography

Batty, M. (1988), 'Informative planning: the intelligent use of information systems in the policy-making process', *Technical Reports in Geo-Information Systems, Computing and Cartography,* 10, Wales and South West Regional Research Laboratory, Cardiff.

Birkin, M., Clarke, G.P., Clarke, M. and Wilson, A.G. (1987), *Geographical information systems and model-based locational analysis: ships in the night or the beginnings of a relationship?,* Working Paper 498, School of Geography, Leeds University.

Breheny, M.J. (1987), 'The context for methods: the constraints of the policy process on the use of quantitative methods', *Environment and Planning* A, 19, pp. 1449-62.

Cook, P., Lewis, S. and Minc, M. (1989), 'Comprehensive transportation models: current developments and future trends', *Journal of the Institute of Transportation Engineers,* June.

Couclelis, H. (1989), 'Geographically informed planning: requirements for planning relevant GIS', Paper presented to the Thirty-sixth North American Meeting of the Regional Science Association. Santa Barbara.

French, S.P. and Wiggins, L.L. (1989), 'Computer adoption and use in California planning agencies: implications for education', *Journal of Planning Education and Research,* 8, pp. 97-107.

Klosterman, R.E. (1989), 'Microcomputers in urban and regional planning: lessons from the past, directions for the future', Paper presented to the International Conference on Computers in Urban Planning and Urban Management, Hong Kong.

Newkirk, R.T. (1987), 'Municipal information systems: challenges and opportunities', *Plan Canada,* 27, pp. 94-100.

Nijkamp, P. and de Jong, W. (1987), 'Training needs in information systems for local and regional development and planning in developing countries', *Regional Development Dialogue,* 8, pp. 72-119.

Openshaw, S., Charlton, M., Wymer, C. and Craft, A. (1987), 'A Mark 1 Geographical Analysis Machine for the automated analysis of point data sets', *International Journal of Geographical Information Systems,* 4, pp. 335-59.

Smith, W. (1989), 'Information society: fact or fiction', in P. Shand and R. Moore (eds), The Association of Geographic Information Yearbook 1989, Taylor & Frances, London.

Taylor, M.A.P. (1989), 'Traffic planning by a "desktop expert"', Paper presented at the International Conference on Computers in Urban Planning and Urban Management, Hong Kong.

Visvalingham, M. (1988a), 'Issues relating to basic spatial units: part 1', *Mapping Awareness,* 2(3), pp. 40-2.

Visvalingham, M. (1988b), 'Issues relating to basic spatial units: part 2', *Mapping Awareness,* 2(4), pp. 42-5.

Worrall, L. (1990a), 'Local and regional information systems for public policy', in M.J. Healey (ed), *Economic activity and land use,* Longman, London.

Worrall, L. (1990b), 'Information systems for urban and regional planning in the UK: a review', *Environment and Planning B* (forthcoming).

Worrall, L. (1990c), 'Improving the quality of "Database UK"', Paper presented at the 1990 Mapping Awareness Conference, Oxford.

Worrall, L. (1990d), 'Issues in the application of GIS in urban and regional policy-making environments', Paper presented to the EGIS '90 Conference, Amsterdam, The Netherlands.

Zevenbergen, M. (1989), 'Spatial Analysis Tool: making ARC/INFO available to the regional planner', Paper presented to the 29th European Congress of the Regional Science Association, Cambridge.

Chapter Two
A Review of Geographic Information Systems applications in Europe

Henk Scholten and Maurits van der Vlugt

Introduction

About a hundred years ago, there was a series of major debates in several Western European countries. The subject of the debates was the right of separate existence of a science of Human Geography. The discussion was protracted but, in the end, the successful defence of human geography was that a specific component — location — was shown to be of elementary significance. The emphasis on analysing and explaining the location of human activity eventually became the foundation for the development of a new science. Since then, there has been a massive growth in the number of analytical concepts in human geography relating to locational analysis: these include, for example, the analysis and explanation of spatial diffusion processes and spatial patterns, spatial interaction modelling, facility location analysis, accessibility measurement and the systematic definition of catchment areas.

After World War II, a science of human geography developed which was largely concerned with the development of spatial statistics and model-building. The movement began in the United States, then emerged in Britain in the late 1960s and by the 1970s it had spread throughout the rest of Western Europe. In conjunction with the development of spatial modelling, the emergence of the computer throughout the 1970s had a major influence on the collection and use of data in the social sciences in general and in human geography in particular. However, despite an overriding concern with the location, it is remarkable that many of the information systems which were developed and used in the 1970s and early 1980s were deficient in the way in which they had incorporated the locational component.

There are two main explanations why these early systems did not and could not efficiently incorporate the locational component. First, a locationally-referenced data set consists of an enormous number of coordinate pairs and the storage and manipulation of these data require powerful computer resources which were not generally available at that time. Second, and of more importance, is the variety of spatial levels at which the data required for complex spatial analysis must be held (these include points, lines, polygons, streets, neighbourhoods, municipalities, regions and countries) — this made it difficult to store data in an efficient way. So, in a period when there was a significant development of spatial statistics and model building, the first geographic information systems were built without a real geographical component. By using mapping software data could at least be displayed, but that was all. At the end of the 1970s, developments in the computer and software industries, particularly in relational data base management systems, provided the opportunity for building 'real' Geographic Information Systems.

In Western Europe, we had to wait until the 1980s for the large-scale development and application of GIS, which had already happened in North America (particularly Canada). In recent years, there has been a massive upsurge in enthusiasm about the development of GIS and many of its proponents have seen GIS as a means of unifying several discrete areas of geographic research and analysis. Others have been more sceptical and have remarked particularly upon the lack of integration of GIS and model-based analysis (see Birkin et al, 1987) — indeed we also argue that model-based analysis is an essential ingredient of a 'real' GIS.

If we are to test whether GIS is achieving its full potential, and to identify how GIS can contribute more effectively to the development of urban and regional policy, we need to explore several issues: we need to identify and review applications in the various fields of interest in GIS and identify distinct groups of users; we need to identify what kinds of applications are being developed to serve different ends; we need to identify and evaluate the actual and potential benefits which result from GIS development and use, and finally, we need to expose the underlying trends in the development of GIS. In this chapter, we will try to explore these issues by reviewing some of the growing number of GIS applications in Western Europe. The review is, of course partial, but nevertheless the discussion of the various applications examined goes some way to exploring these issues.

This chapter contains five main sections. First, we develop a typology of what we call 'Spatial Information Systems' in which to define and locate GIS; second, we classify GIS users into a fourfold typology and examine their distinctive needs; third, we review several discrete application areas which have provided a foundation for the development of GIS ; fourth, we review a variety of GIS applications drawn primarily from Western Europe before attempting to identify trends in, and future directions for, the development of GIS in the final section.

GIS within a typology of spatial information systems

In almost every GIS journal, a lengthy discussion has taken place in an attempt to define the term 'GIS'. Many labels have been used as synonyms for GIS: these include geo-base information system, geo-data system, spatial information system, geographic data system, land information system, natural resource information system, multi-purpose cadastre and multi-purpose land information system (see Clarke, 1986; Burrough, 1986; Parker, 1988). This terminological confusion can be explained by the fact that GIS is a young science which is related to many other disciplines and fields of technological development all of which deal with spatial data handling. The related areas include as disparate subjects as remote sensing, regional economics, cartography, surveying, geodesy, photogrammetry and, of course, geography.

There are many definitions of a GIS (see Marble and Peuquet, 1983;

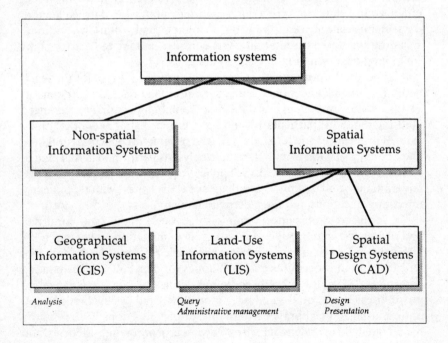

Figure 2.1: *A classification of (Spatial) Information Systems*

Burrough, 1986; Calkins and Tomlinson, 1984; Berry, 1986) and in Figure 2.1 we attempt to develop both our definition of a GIS and to locate GIS within a generic class of Spatial Information Systems which, we argue, can be divided into three subclasses — Geographic Information Systems, Land Information Systems and Spatial Design Systems. Unlike Rhind and Mounsey (1989) we argue that there are essential differences between GIS and LIS and that, while in practice the distinctions between the three types are not as rigid as we may imply here, they are significant.

In general, the term Computer Aided Design (CAD) refers to graphical systems which support the work of architectural or industrial designers. CAD provides the facilities to perform calculations (volume, weight, amount of materials) and to produce technical drawings and three-dimensional displays of designs. A spatial CAD is a tool which can be used for urban and landscape design and it consists of two main elements: a data base containing spatially referenced (two or three-dimensional) data and procedures and techniques for data collection, data manipulation (e.g. viewing, zooming and rotation), editing, visualisation and presentation. We consider automated cartography (mapping) as, essentially, a CAD system.

A Land-use Information System (LIS) is primarily a tool for the legal, administrative and economic management of land resources. It consists of two main components: a data base containing spatially referenced land-related data (particularly land parcels and spatial networks) and a suite of procedures and techniques for the systematic collection, updating and querying of data. Minimal spatial analysis is performed in this kind of system. The majority of LIS applications can be found in local government and public utilities such as the water industry, telecommunications and the electricity industry.

We see a geographic information system as essentially a tool for urban and regional research, policy analysis, policy simulation and planning. A GIS consists of a data base containing spatially referenced data and, like an LIS, a set of procedures and techniques for data collection, updating and query. In addition to these basic facilities, a GIS should contain procedures for spatial analysis, modelling, policy evaluation and optimisation and elaborate cartographic display functions. GIS research is a unique field of endeavour with its own sets of problems, 'theories' and techniques for finding solutions to problems which manifest themselves spatially — though the central focus of a GIS is the manipulation and analysis of spatial data.

The differences between GIS and LIS are not great when viewed from a hardware and software perspective. The difference lies in the fact that

LIS are used primarily for the storage and retrieval of spatial data while GIS are used essentially for more complex spatial analysis. Not surprisingly, the functionality of an LIS is considerably below that of a GIS. Another difference between LIS and GIS is the scale and precision of the spatial data held: LIS usually operate on a fine spatial scale (perhaps 1:500 or 1:1000) while GIS data are often on a scale of 1:10000 or smaller (though there are exceptions to this rule of thumb which are discussed later).

The differences between GIS/LIS and CAD are much more radical than the differences between GIS and LIS. The orientation in CAD is mainly on visualisation and this has necessitated the development of fundamentally different data structures and approaches to data manipulation. In our view, the differences between CAD and GIS/LIS are too large for CAD to find independent applications in spatial planning. Examples drawn from the aircraft and automobile industries show that the benefits of CAD can be enhanced by adding related data such as wind patterns and that it is essential to augment basic CAD applications in this way if they are to be of use to physical planning since judgements about the quality of the future environment must be based on both qualitative (design) and quantitative (analytical) information.

The integration of CAD into GIS/LIS is important but, at the current stage of development, it has largely been limited to the physical exchange of data from one system to the other — this is only a first step towards the more effective integration of both types of system. An example of this can be found in the city of Utrecht in the Netherlands, where the city planning agency has experimented with a CAD system which is used in the planning and design of public space and building locations. To use the system effectively, administrative and topographical data had to be included from the city's existing geographic information systems. The conclusions drawn from this experiment were that the use of such a spatial CAD system was economically worthwhile and feasible both from a technical and from an organisational perspective (Hitzert, 1989).

A typology of GIS users and an analysis of their needs

Everyday practice has shown that a number of common problems can be distinguished in using GIS (and LIS) and that these problems are closely related to differences in the type of information needed by different types of user. Within an organisation there are many different tasks, and increasing functional specialisation over the last few years has meant that the information needs of decision-makers, for example, are not the same as the information needs of policy analysts. For decision-makers, the nature of their role means that it is usually a matter of evaluating different policy options without referring to the large volume of less directly applicable data on which those policy options are based. Essentially, their overriding need is for information at a strategic level which will enable them to evaluate between the various policy options set before them. It is the task of the policy analyst to prepare policy options for decision-makers and to supply strategic-level information based upon a rigorous analysis of usually large volumes of data. To achieve this, the policy analyst has to research a given policy field, to bring the key issues into focus, to identify the options and to assess their redistributive or selective effects. While this latter series of tasks has to take place whether advanced statistical or GIS techniques are available or not, we argue that the tasks will be performed more efficiently, more rigorously and more effectively using GIS.

It is clear that different types of users place differing demands on GIS and by exploiting this user-needs approach, we believe that we can define more rigorously the types of spatial analysis required from a GIS and explore more fully the criteria which a GIS must meet for it to be effective in a wide range of practical policy-analytic situations. We have identified four main groups of users — all of which have specific requirements for spatial information. The four groups, together with a brief summary of their distinctive spatial information needs and their demands are presented in Figure 2.2 (after Scholten and Padding, 1990). The model is perhaps most applicable to the public sector, but, with some minor modifications, it can be made more generally applicable. The various information needs are discussed in more detail below.

User type 1: information specialists (the professional user)

This group comprises users who regularly work with computer software though the emphasis within this group tends to be more with technical, methodological and developmental issues than with applications in practical decision-making environments. The main problem focus here is with the application of advanced statistical and mathematical methods and techniques to the processing of data material. Despite the analytical potential of GIS, it would appear that the functionality of the present generation of GIS systems is limited in this field.

The heavy emphasis on analytical techniques means that a large amount of work carried out by this group has been focused explicitly on the locational component (e.g. overlay analysis, network analysis, buffering). In addition, applications have addressed issues like

TYPE OF USER	INFORMATION DEMAND	USER DEMAND	TYPE OF GIS	DEVELOPMENT
A: Information specialist	Raw data	Analysis Flexibility	Large Flexible	Links to other packages
B: Policy analyst	Raw data and pre-treated data (=information)	Analysis Good accessibilty	Compact Manageable	Macro Languages Interfaces to other packages
C: Policy decision-maker	Strategic information	Good accessibility to users Weighting and optimalisation models	Small and concise	User friendly interface Key Information
D: Interested citizen Special interest groups	Information	Good accessibility to users	Small and concise	User friendly interface

Figure 2.2 *GIS, type of user, kind of need (after Scholten and Padding, 1990)*

identifying optimal locations for certain industries, examining the consequences of population growth and development for local amenity structures and evaluating wide-ranging planning objectives from the perspective of conflicting interest groups. The problem with most of the GIS developments in these fields is that they have not been extended far enough and they need to be supplemented by the addition of complementary geographical analysis techniques (Openshaw *et al.*, 1987). In addition, the types of analysis referred to above require the construction of complex analytical models which in turn have implications for the way that data is structured if analysis is to be performed in a reasonable amount of time. Developments such as quadtrees (see Shaffer *et el.*, 1989) mean that the complex analytical models within GIS can be executed in a reasonable time-scale on PCs though further developments are needed in this area.

User type 2: policy analysts

The people who prepare policy studies often work closely with information specialists, though as software has improved and technology has become more powerful and more available, the distinction between these two groups may be becoming less pronounced. Instead of specialising in the arena of methods and techniques, policy analysts tend to be more involved with policy preparation in a specific area of spatial planning. In the plan-making process, it is important to be able to make use of the most up-to-date analytical methods and, up to now, this has only been possible through cooperation with information specialists.

For many standard activities, an extensive knowledge of the GIS command language is required, and often this group of users does not have the necessary operational knowledge. However, actual practice shows that many applications involve only a small subset of the extensive functionality that GIS offers — examples of the more simplistic use of a GIS include rudimentary overlay analysis, buffering or data display. Such activities can be standardised through the use of macros which can execute simply and efficiently the most frequently used operations via a menu-driven environment. In LIS applications this aspect is especially well catered for by the design and programming of a wide range of standard procedures.

User type 3: policy decision-makers

This group is, at the moment, largely confined to the public sector and comprises two distinct groups: senior executives and politicians. The main task of these two groups is the evaluation of policy options on the basis of information supplied by policy analysts. Up to now, the development and application of automated decision-support systems for these users has been virtually non-existent. The increasing complexity and amount of available spatial data means that there is an increasing need for a decision-supporting environment for these users. A necessary condition for the creation of this environment is that the information and intelligence supplied to decision-makers and the rules for evaluating alternatives should be uncomplicated, unambiguous and easily understood as policy decision-makers have, in general, little affinity with, or knowledge of, automation, statistical methods and mathematical techniques.

User type 4: the interested citizen and pressure groups

This group comprises interested citizens, pressure groups, special interest groups and various social (often locality-based) organisations. The members of this group often like to be kept continually informed about topics such as physical planning policy in general or specific environmental issues in particular. We believe that there is a large potential demand for information systems in which the most important up-to-date information on physical planning and environmental issues is stored. The RIA project in the Netherlands (Scholten and Meijer, 1989) and the the Domesday project in the United Kingdom are examples of the development of spatial information systems designed for widespread public use (Openshaw *et al.*, 1986). It is in this field that GIS could contribute very effectively to enhancing the democratisation of decision-making.

The four groups of actual and potential GIS users we have identified above are geared primarily towards various aspects of government but the typology can be easily extended to other non-governmental organisations. In almost every organisation where there is a need to use spatial information, the ability to interrogate that information will not be equally distributed throughout the organisation. In terms of the further development of GIS, the largest potential user-base will be found amongst those who are currently unable to work independently with the present generation of GIS software. For this group, the opportunity to work with, and to exploit more fully, the potential benefits of GIS, will only arise through the further development of user-friendly interfaces. Apart from interested citizens and special interest groups, the three other user groups we have identified (information specialists, policy analysts and policy decision-makers) are numerically quite small but very powerful in terms of their actual and potential influence on public policy-making and resource allocation.

GIS applications: an analysis by sector

In Europe, until quite recently, GIS (and particularly LIS) applications had tended to take place within a small number of discrete application areas. Below, we will give a short description of these different sectors and try to assess their significance for, and contribution to, the development of GIS.

Land and property information systems

Land and property information is probably the most well developed field for spatial information systems applications. It has attracted a lot of attention from public authorities at various levels of government throughout Europe. This is not surprising given that 'all authorities maintain some form of property based information system to support property management activities' (DOE, 1987, p.151). These systems are difficult to categorise because the term 'land and property information system' has been applied to a wide variety of systems in use by local authorities. Research into the use of these information systems in England and the Netherlands has been limited and most effort has been concentrated on aspects of the design of these systems and on the nature of front-end query systems to interrogate data (Grimshaw, 1988; Bogaerts, 1987). Within LIS, a high level of accuracy and the completeness of the data in any system are essential if the systems are to achieve their full potential and be most cost-effective.

Public utilities planning and management

The telephone, electricity and gas utilities tend to be located within the public sector in most European countries and this has led to the development of a small number of very large nation-serving enterprises. Within each of these three industries there is a prevailing emphasis on the management of large distributive networks and substantial infrastructure. In terms of spatial information systems development, it is useful to distinguish between applications at both the large and small scale. At the large scale (high resolution), applications include monitoring and planning the layout of pipelines and cable networks and the location of facilities. At the small scale (low resolution), applications include the planning of facilities and transmission lines to minimise economic, social and environmental costs.

In Europe, GIS developments in this sector have begun only recently but are expanding rapidly. GIS systems within the public utilities provide facilities for the more efficient management of large distributive networks and, more important, provide a framework for the better integration of relevant information derived from several sources. Before GIS became more generally available, all locational, administrative and network data were stored separately, but through the medium of GIS these can now be easily combined.

Transport, facility and distribution planning

Many of the first computer applications in the spatial sciences took place in this field of research and, as such, an involvement with computing has evolved which has not traditionally used GIS techniques. As a result of this, GIS applications are only just beginning to be developed in this field. Most practitioners in this sector do use data base management systems and statistical packages in combination with mapping packages to display their results but given our earlier definition of a GIS, this falls short of being a real GIS.

The factor which most applications in this field have in common is that the locational component is usually at a national scale and that, usually, only a limited number of locational characteristics are taken into account. The fact that the locational component in these studies is not as predominant, nor at as fine a scale as in other application areas, has meant that the use of 'traditional' analytical methods still dominates the field. The potential for integrating GIS within this application area is very promising. Apart from data storage and graphical display, GIS has the ability to make operational certain specific locational concepts and techniques: these concepts include the systematic analysis of proximity, accessibility, connectivity and density. In addition, GIS can also perform traditional locational analysis. The effective integration of all these functions and facilities would be very difficult to achieve, if not impossible, using traditional (i.e. non-GIS) methods.

Regional and physical planning and environmental protection

The potential for the use of GIS here is much more obvious than in the application areas discussed above. This is because all the data and analysis concern specific geographic regions and so all the relevant information can be easily stored, combined and manipulated in a GIS. It is disappointing to note that applications in this area are not particularly well developed except perhaps in the Netherlands. In the Netherlands in 1985, the National Physical Planning Agency was the only large GIS user but since then many more GIS users have emerged (for example, the Ministries of Agriculture and Environment which began to develop large GIS applications in 1988). In the other Western European countries we can see signs of the same 'GIS-boom' but it must be emphasised that the practical use of GIS is much more advanced in fields such as utilities management and natural resource management than it is in urban and regional planning.

Academic research

In both Great Britain and the Netherlands there is a very strong platform for GIS research at several universities. In Great Britain, the Economic and Social Research Council (ESRC) has allocated one and a half million pounds over three years for the implementation of a Regional Research Laboratory Initiative and this has recently been augmented by the establishment of an Urban Research Laboratory jointly based at Liverpool and Manchester Universities. This is a major initiative designed to promote the integration and exploration of several forms of social, economic and environmental data (Masser and Blakemore, 1988).

In Great Britain, there are currently eight 'regional' laboratories and one
'urban' laboratory. The centres have four main functions: data
management, software development, spatial analysis and research
training.

Similar centres have been developed in the USA (the NCGIA —
National Center for Geographical Information — see Abler, 1987) and in
the Netherlands (NexpRI — see Ottens, 1988). The NexpRI initiative in
the Netherlands was funded through the Dutch National Science
Foundation and was opened at the University of Utrecht in 1989.
According to Ottens (1988), the aims of the NexpRI initiative are to
provide on-line information on geographic information processing, to
bring together the demand for, and supply of, GIS services and support,
to organise and promote computer assisted education in geographic
information processing and to function as both a platform for GIS users
and developers in the Netherlands and as a focal point for international
cooperation.

A basic theme which underpins these developments was identified by
the Chorley Committee (DOE, 1987) which pointed out that a
fundamental change was taking place in the routine use of geographic
data and that this would have far-reaching consequences for a wide range
of users. While they saw the development of technology as crucial, they
saw that the most effective means of achieving the full potential of GIS
was by improving both the level of public awareness about GIS and the
quality of GIS skills training.

The application sectors we have reviewed above have made, and are
continuing to make, a contribution to the further development of GIS. In
practical management applications, such as land management and public
utilities management, the benefits of spatial information systems,
whether they be LIS or spatial CAD applications, can be seen to be
making a very tangible contribution to the more effective planning and
management of resources because they are designed for very specific
purposes. The development of general purpose GIS is much harder to
evaluate and this may account for the lower level of development in urban
and regional planning applications. Clearly, there is a considerable
education task to be undertaken to convince decision-makers of the real
benefits of GIS, and the awareness-building task forms one of the central
roles of the academically based GIS centres in the USA, Great Britain
and the Netherlands.

A review of GIS and LIS applications in Europe

Having discussed in general terms the opportunities that Geographic Information Systems offer and the areas in which they can be, and have been applied, it remains to examine in more detail various examples of the application of GIS in Europe. Here the objective has been to select and describe applications of GIS/LIS which have produced superior results than were possible before, or have produced results more efficiently than before or have produced results which were unachievable without GIS methods.

Land Information Systems

As stated above, LIS applications tend to be geared to the storage of data in large data bases and to the relatively simple interrogation of data: they are not, by definition, geared to performing complex spatial analyses. What LIS has to offer compared to 'traditional' data bases is, of course, the spatial component, but this is not only concerned with the location of various entities but also with the ability to perform accurate administrative calculations such as measuring land parcel areas or boundary lengths often as an input to taxation assessment. The emphasis in the use of LIS as opposed to GIS lies, therefore, on supporting routine administrative tasks (Grimshaw, 1988) and the combination and integration of the different information systems.

Although applications in this field are emerging in a piecemeal way all over Europe, often in the form of a cadastral LIS used by local authorities, more integrated applications on a national level are at a more advanced level of development in Scandinavia. For example, a major project was started in Finland in 1985 in which around thirty state organisations, several municipalities and several private organisations became involved in an exercise to collect spatial data for joint use. LIS technology has, in this instance, made a large spatial data set available to many users in an efficient way (Ahonen and Rainio, 1988). Also in Scandinavia, the Norwegian land accounting system has been developed to provide access to information on actual land-use, on land-use change, on plans for future land-use, and on the suitability of land for different forms of land-use. It must be pointed out that the Norwegian system has been developed primarily to serve the national land-use planning authorities (Engebretsen, 1987).

In the Swedish National Land Survey, research has been carried out to develop a general system for cadastral surveys culminating in the development of a land data bank for use by local authorities in land-use planning and in land consolidation (Falk, 1987; Piscator, 1986). Also in Sweden, two national elevation data bases have been developed. In both these data bases, elevations are stored for intersections of regular grids based on grid sizes of 25km × 25km and 5km × 5km and as such are an exception to the general rule identified earlier that LIS data bases are of a large scale. A comprehensive overview of these data bases is given in Ottoson and Gut (1986). In the Netherlands, there are two large mapping projects in operation. The first of these projects aims at the development of a map of the Netherlands at a scale of 1:1,000. The project started in 1975 and currently maps of about 25 per cent of the land area of The Netherlands are available in digital form (Bogaerts, 1987). The second project aims at the development of a street segment oriented system for the whole of the country. This project is in the hands of a private company (Tele Atlas) which uses it commercially as a base for a large variety of spatial information systems.

Forestry

In Canada, forestry was the original application area of GIS (see Tomlinson, 1987, p.209) though pure applications in this field in Europe are relatively unimportant because the amount of forest cover in Europe is much smaller than it is in Canada. A common theme which underlies GIS applications in forestry in Western Europe has been the need to examine the extent to which forestry might compete economically with alternative land-uses. In Western Europe, major land-use changes are taking place as a result of EEC farm policies: one of the more profound impacts of these policies will be the afforestation of land that is currently used for agricultural purposes (Padding and Scholten, 1988; Bunce *et al.*, 1989). The afforestation of land has proved to be a politically contentious issue but the advantage of GIS applications in this area is that it can provide relatively objective criteria and methodologies for the selection of land for afforestation.

In Austria, GIS has been used to depict the distribution of those spatial factors which are responsible for causing forest damage. Several models have been developed to analyse, predict and explain the dynamic development of forest damage. The GIS in this application area serves as a unified data base for the storage of a large amount of data from different sources, and subsequently for the storage and display of the results from the analysis undertaken. The main output from this system takes the form of a time-series of maps charting the dynamics and evolution of forest damage (Grossman, 1987).

A final example of the application of GIS in the forestry sector is described by Horn and McLaren (1988). In forest management and design, visual quality is seen as crucial to the success of a design when assessed for environmental impact. The traditional design process has improved through the use of computerised mapping and visualisation techniques. In this way, the visual result of a proposed forest design can be previewed from a natural, three-dimensional perspective which includes relief and visible artefacts as part of the design. Such applications could also qualify for a place within the group of CAD-applications.

Transportation planning and research

In Great Britain, a GIS application has been developed by the South-East Regional Research Laboratory in collaboration with British Rail (Whitehead, 1988). The objective of this project is to assess the impact of the changing geography of the South-east region, in particular the redistribution of population, the changing structure and location of employment and the growth of private car ownership on travel demand. The first stage of the project was to examine the impact of the changing distribution of population in relation to the nine hundred railway stations in the rail network of South-east England and to identify areas of significant population growth not currently served by the existing rail network (Whitehead, 1988). In this study, traditional socio-economic research methods have been combined with GIS methods. In the Netherlands the planning of a new route for the TGV (High Speed Train) was undertaken using GIS techniques. In this application, an experimental model for calculating the financial and environmental cost of the planned route was implemented in the GIS (Osinga, 1988).

Health care planning and epidemiological research

Many studies and preliminary investigations have suggested that mortality and other health indicators are distributed unequally over space. Arising from these observations, several researchers have sought to develop computerised information systems to assist in the explanation of spatial patterns in health inequalities and various related statistics. Recently, an entirely new technique for the automated analysis of medical data has been developed by building what is known as a Geographical Analysis Machine (GAM) (see Openshaw *et al.*, 1987; Gatrell, 1988). This new methodology has been described as constituting a quantum leap in epidemiological research which will have many potential applications in other areas of spatial statistics once it has been perfected. The GAM offers a radically new, fully automated and spatially systematic approach to the search for spatial clusters of cancer occurrences. It confirmed the existence of a well-known acute lymphoblastic leukaemia cluster in Northern England, but far more surprising was the 'discovery' of another even larger hitherto unknown cluster (Openshaw *et al.*, 1987). The benefits of the use of GIS here are obvious.

Site evaluation systems and market analysis

The development of commercially orientated GIS in this area fall into two distinct categories: site evaluation (for which there is a long history of traditional (non-GIS) applications) and market segmentation analysis. The evaluation of the retail potential of individual locations and the ability to compare the retail potential of various sites is an obvious subject area for the development and use of GIS applications and developments in this field have much in common with general developments in facility location analysis described elsewhere.

Within the development of target marketing systems, vendors of a product identify a consumer group they think will want to buy their product and define that group in terms of its socio-demographic characteristics. By matching the assumed characteristics of the target consumer group with, for example, census data, it is possible to segment the initial selling area into a number of smaller areas where it is hoped the highest levels of response to either direct mail or some other form of advertising will be achieved. The process of target group selling and market segmentation and its links to GIS can be seen in several studies (see Beaumont, 1989; Openshaw, 1987; Clarke, 1990). This is one of the few real commercial applications of GIS.

Environmental protection, risk management and emergency planning

Several specific applications in this field are related to risk analysis and risk assessment with reference to potentially hazardous facilities or occurrences. Different departments within various organisations need spatially referenced information to assist them in their planning and decision-making in the event of a major environmental threat or disaster. Such a disaster might originate at a point source such as a chemical explosion at a factory followed by the emission and spread of a toxic gas cloud (e.g. Chernobyl), or it might be a 'linear' risk emerging from the transportation of a dangerous gas by rail, or it could be an areal hazard such as a major flood.

Effective emergency planning requires information about the likely dispersal of the hazard in space and through time, the distribution of the population at risk from the incident (under various assumptions about

dispersal) and the location of resources needed to tackle the problem
(e.g. availability of transport and temporary shelter). Specific
applications in emergency planning have been designed to counter the
environmental risk of chemical industries (Fedra, 1986), for planning
evacuation in case of flooding (Southworth and Chin, 1987), for the
simulation and mapping of air pollution (Koussoulakou, 1988) and for
the management of disasters in the North Sea (TNO, 1988). A GIS
provides an effective and efficient medium in which to gather, combine,
analyse and display information from several sources.

In Britain, GIS has been used experimentally to select disposal sites for
radioactive waste. The research problem was not to select the 'right'
solution to the waste problem, but to select what were technically the
most suitable sites. GIS helped to improve the selection procedure and
also made the final selection more justifiable as it was chosen on 'hard'
criteria after an exhaustive search (Carver, 1988). The danger of
justification by saying 'the computer said so' is very much present here,
but this applies to all computer use in policy-analytical applications, not
just to GIS.

Environmental planning and impact assessment

The number of GIS applications found in this field in Europe is high and
this is because of the strong locational aspect in these studies and because
of the political sensitivity of many of these studies. The way that GIS fits
into planning organisations and planning processes has been the subject
of several papers (see Scholten, 1987; Scholten and van der Velde, 1987;
Batty, 1988) and we hope to elaborate further on this debate by reference
to the applications we describe below.

Environmental impact studies and ecological balancing methods for
land-use planning and land consolidation are gaining more importance in
the planning process. The ever increasing competition for space between
different activities is generating, and will continue to generate, more and
more land-use conflicts. GIS is a useful tool for making an inventory of
present and planned land-use, assessing environmental quality and
assessing the likely impact of proposed developments and land-use
change. Several studies on this theme have been developed in West
Germany and Britain (see Schaller, 1986, 1987; Schaller and Haber,
1988; and Green, 1988). Special mention must be made here of two
particular projects. The first is a large joint European project called the
CORINE project, which is being set up by the Commission of the
European Community. This broad-scale GIS has been designed to bring
together information on a wide range of topics as part of a Community-

wide environmental data base, and, when complete, will also provide the foundation for the development of a wide range of GIS procedures and applications (Briggs, 1988). The second large-scale GIS exercise is the MEDASE project which is being established by several nations bordering the Mediterranean: GIS will play a central role in this project which is being designed to improve the information base for environmental management (Montanari, 1989).

In the Netherlands, GIS techniques have been used in several different capacities in the national physical planning process. The *Fourth Memorandum on National Planning* was published in 1988 and this sets out the government's broad policies for national planning until the year 2015. Within this process, GIS has been used for a number of purposes: first, to estimate the total cost of implementing these plans (see Groen,1988); second, to investigate the feasibility and impact of building one million new dwellings in the Randstad area (see Geertman and Toppen in this volume) and also to develop regional environmental regulations. Perhaps the most publicly visible and tangible development to emerge from the process was the creation of an overall spatial information system which was developed to inform the public about the Fourth Memorandum — this is the so-called RIA project (Scholten and Meijer, 1989).

From our review of GIS applications, we can see that the range of application areas is very wide and that GIS techniques enable the policy analyst and planner both to undertake tasks that they have been unable to perform using more traditional methods and also to undertake many of these traditional tasks more efficiently and effectively than before. The improvements in analytical potential provided by GIS will enable planners and policy makers to have a greater impact on public policy formulation, evaluation and implementation than they have had before. We see the main advantages from GIS to be in the following areas:

1.   in the better integration of information from a wide range of sources;

2.   in the more effective presentation of information in a form which is more comprehensible to decision-makers who may not be able to distil knowledge from pages of tabular output;

3.   in the ability to simulate the impact of policy choices in politically sensitive policy areas and geographic localities;

4.   in the ability to simulate the impact of disasters and hazardous events; and

5.   in the ability to assemble a comprehensive set of socially and

politically relevant information and to build this into an accessible system for public use.

Clearly, GIS is still in its infancy and many interesting and socially useful applications remain to be developed: it is to directions for the future that we turn in the next section.

Directions for the future development of GIS

GIS: a unifying framework for social science and policy analysis

It is very difficult to write a paper about the state of the art of a young science as the rapid pace of development means that much of information is outdated before it can be published. The aim of this paper, therefore, has not been to present a state-of-the-art review, but rather to give an overview of GIS applications in Europe. This also explains why we have chosen not to end our review with a traditional conclusion but rather to try and identify the underlying trends in the development of GIS and to attempt to predict the direction in which GIS will develop in years to come.

A major emerging trend we perceive is the emergence of a more unified and more interdisciplinary approach to research and policy analysis particularly, though not exclusively, in the social sciences: this has been brought about by the development of GIS. It may be argued that GIS, through its potential for storing and integrating large amounts of spatially-referenced data from diverse sources, should develop into a tool for the synthesis of many fields of research in addition to the role as an analytical tool which it fulfils at the moment (Abler, 1987, p.323). Linked to this point is the observation that GIS research is necessarily and essentially multi-disciplinary in nature. Geologists, geographers, cartographers, economists, social scientists and planners may work together on a single project. If GIS is to develop as a powerful tool, the barriers between the various disciplines of users will need to be broken down and they will need to develop a common language (Rhind, 1988, p.27; Birkin *et al.*, 1987, p.19). In current practice, however, this is seldom the case and research teams are usually formed within a single discipline. By offering a unified data base GIS will eventually help to break down these barriers — the epidemiological research cited above (Openshaw, 1987) is a good example of the more harmonious working of the medical profession, geographers, policy analysts and computer professionals.

While we are of the opinion that GIS provide an effective means of assisting the planning process and assisting in the solution of particular problems, it is important that GIS should be focused on practical planning problems. As Abler (1987, p.322) remarked 'I am uncomfortable with the degree to which GIS enthusiasts are committed to providing answers to questions nobody may ask'. This, we believe, reinforces the need to promote awareness about the tangible benefits that GIS can offer among policy analysts (i.e. User type 2) and policy decision-makers (User type 3). At the moment, GIS is too much the preserve of the 'GIS enthusiasts' who tend to populate User type 1 — the 'information specialists' who work with GIS daily, though not necessarily applying the resources available to the areas of highest social priority.

The potential that GIS offers is not well enough understood among possible users and therefore not enough questions are being generated from the strategic level within organisations: educating decision-makers to ask the right questions is essential if the information specialists are to focus their research more concisely. Consequently, there is a tendency for GIS specialists to provide accurate answers to irrelevant or low priority questions. More research into the range of questions that GIS should be helping us to answer is desperately needed and these questions are only likely to be articulated at the strategic policy-making level within organisations. The essential message is that 'GIS is not an aim in itself, it is a means to an end' (Scholten and Padding, 1990, p.1) and GIS will be most effective if it is targeted towards solving those problems of high social and political priority.

GIS development streams

Future developments in GIS can be divided into two main streams. The first stream has a technological and methodological focus and is largely concerned with the further development and improvement of currently used techniques for complex spatial analysis, visualisation and data storage. Several research agendas have been compiled in this area with various private companies developing predominantly conceptual frameworks for GIS applications. The second stream is more focused on the identification of new application areas and the introduction of new techniques within GIS. The second stream is scientifically more interesting but specific research themes are harder to pinpoint, though we have tried to assemble an inventory of specific fields in which future GIS development is desirable or can be expected.

The first stream of future developments will involve the extension and maturing of currently used techniques — in other words, the emphasis will be on improving the things which we can already do. Improving the visual output from GIS is a key priority: GIS mainly produces maps which, because they are not usually displayed on paper but on computer screens, can be made dynamic to display development over time (the forestry example cited earlier is a good example of this — see Grossman, 1987). The shapes, sizes and colours of map elements are also relatively easy to alter to improve visualisation (Fedra *et al.*, 1987). Another part of the display aspect is the production of three-dimensional maps which are an effective means of presenting information and are already becoming widespread in GIS. However, they take up a lot of computer memory and processing time: the increased use of 3D maps will necessitate improvements in the power and sophistication of the hardware available.

Another area for development within the first stream is the need to improve GIS linkages to other geographical techniques and models. For many years, geographical model builders and GIS users have been living separate lives even though they have a lot to offer each other. One can think of the incorporation of spatial interaction models, network analysis, spatial diffusion modelling or space-time analysis in GIS. Birkin *et al.*, (1987, p.19) argue that this requires the advanced interfacing of GIS techniques and modelling software, though in an ideal situation the GIS would act as a 'shell' to which other software could be linked.

A final development area concerns improving data structures. With the growing complexity of operations using spatial data, we are going to need more and faster access to this data. Partly this is a hardware problem, with which we are not concerned here. For the other part, however, research needs to be directed towards developing faster and more efficient data structures to store and retrieve geographical data. The 'quadtree concept' is a good example of how improvements in data structures can radically improve the effectiveness of GIS in practical planning environments (Shaffer *et al.*, 1989).

Although many of today's techniques have not yet reached maturity, GIS users are setting their sights still higher and are searching both for other problems that GIS can address and for more advanced techniques with which to address these problems. One problem with GIS is that it can only deal with 'hard' data. Areas can only be defined using hard boundaries when in reality softer or 'fuzzy' boundaries would be more appropriate (a good example of a fuzzy boundary is one which defines a functional urban region). Within current GIS applications, such phenomena have been considered as sharply defined largely because the techniques needed to handle the problem in a different way did not exist. Therefore, research is needed in the field of 'fuzzy data' and 'fuzzy logic' in order to answer the question of how 'fuzzy' spatial entities can be described with digital data and how can they be manipulated. It is in this field that GIS can make a fundamentally unique contribution to developments in spatial analysis. Other new developments in spatial analysis where GIS can play an important role are in the analysis of point patterns, in the development of 'pointillistic' geography where data are individually stored instead of being aggregated into some arbitrary statistical area and in the further development of a theory of spatial relations (see Abler, 1987. pp.305-6).

Another new research theme in GIS is error analysis. All maps and all data are, in one way or another, inaccurate and little is known about the behaviour of these errors when maps of different scales, or numerical data from different sources, are combined. Some of the questions which may arise are: what percentage of error is acceptable, how do errors evolve during data processing, what is their effect on the final outcome of any analytical process and how can we develop and display a measure of reliability on the resulting maps? These issues are still very much in uncharted waters (Charlton *et el.*, 1989).

The potential that GIS offers can be greatly enhanced when techniques drawn from the field of Artificial Intelligence (AI) are incorporated. AI techniques could effectively be applied within GIS for a number of purposes where human expertise is currently required. Possible applications of AI techniques within GIS are in the following areas:

1. Automated data entry: getting maps into computers is a time consuming business, and while scanning is much faster than digitising, there needs to be a lot of editing afterwards to restore scanning errors and to build topological relationships. 'Intelligent' scanners would be of considerable help here.

2. Map interpretation: this involves, among other activities, error checking, generalisation and feature extraction. In the future, expert systems combined with a pattern recognition routine, could take over many of these tasks.

3. Producing cartographic output: a lot of 'human' work is involved, for example in designing legends, choosing textures, placing text. Intelligent systems will eventually take over this type of work.

4. The creation of fully 'intelligent' GIS: a present-day GIS will do just as it is told and nothing more or less. It will not question the validity of the required operation or remember whether or not it has previously been used to answer the same or a similar question.

Within an intelligent GIS several facilities would need to exist:

1. A geographical knowledge base: this would be used to check whether the operation requested by the user was methodologically correct. If it was not correct, or if the interpretation of the results was likely to produce an erroneous image of reality, a warning would be given. The knowledge base could also be used to suggest the most appropriate methods to tackle a certain problem.

2. Computer-based teaching: an intelligent GIS should be able interactively to teach the user how to work with the system and tell the user where he had gone wrong and why.

3. Answering 'What if' questions and self-learning: eventually, it should be possible to ask general questions of the system, such as 'what happens to traffic flows if employment in a given city rises by 25 per cent?' The system will try to find a general solution path for this problem. If it is not available or incomplete, it will ask the user direct questions on how to solve the problem. It will, of course, use the above-mentioned geographical knowledge base and after producing a result, the system will store the procedure used and the information gained from the user for future use. We realise that this kind of use of GIS is perhaps some years away, but this is a final goal towards which we can work. Some of the intermediate stages towards this goal are currently feasible (see Fedra et al., 1987; and Openshaw et al., 1987).

Emerging from our discussion of intelligent GIS, we will briefly examine the field of Decision Support Systems (DSS). The difference between an information system and a DSS lies in the fact that in a normal information system the procedure to compile information out of raw data is relatively fixed. The available information is prearranged and the user has little or no influence on the final outcome. In a DSS, however, the user is expected to employ his own expert judgement in compiling and producing information from the available data. In this way, the user can examine, for example, several policy alternatives, can decide to use a certain model and can study the effects of changing certain parameters of a model. A DSS is designed to help a policy maker make decisions founded on more useful and 'made-to-measure' information than he can obtain from a traditional information system.

When methods of spatial analysis and intelligent techniques are built into GIS, it automatically evolves into a geographical DSS. Therefore, we think that the use of GIS in the development of DSS is one of the most important applications of GIS that will emerge in the near future. This development needs to be seen in the context of building smaller and more manageable GIS-applications for users other than the group of information specialists we have identified.

There is no question about the fact that GIS is the fastest growing application of information technology in the social sciences. Apart from the more detailed aspects discussed above there are some general aspects we have to be aware of in the near future. First, a basic familiarity of what GIS has to offer should be part of the training of all scientists who are involved with spatial data. Second, the systematic geo-coding of large-scale data sets will provide a basis for integrating data which have been until now incompatible or fragmented. The CORINE project (Briggs, 1988) is an example of how difficult it is to start with this kind of project, but on the other hand it shows how large the opportunities are and the potential that GIS has for improving the quality of urban and regional policy analysis.

Bibliography

Abler, R.F. (1987), 'The National Science Foundation Center for Geographic Information and Analysis', *International Journal of Geographical Information Systems,* 4, pp.303-26.

Ahonen, P. and Rainio, A. (1988), 'Developing a query system for joint use of spatial data in Finland' in J.C. Muller (ed), *Environmental applications of digital mapping,* Proceedings of Eurocarto Seven, ITC, Enschede, The Netherlands, pp.77-86.

Batty, M. (1988), 'Informative planning: the intelligent use of information systems in the policy-making process', *Technical Reports in Geo-Information Systems, Computing and Cartography,* 10.

Beaumont, J.R. (1989), 'Market analysis, commentary', *Environment and Planning A,* 21, pp.567-70.

Berry, J.K. (1986), 'Learning computer assisted map analysis', *Journal of Forestry*, pp.39-43.

Birkin, M., Clarke, G.P., Clarke, M. and Wilson, A.G. (1987), *Geographical information systems and model-based locational analysis: ships in the night or the beginnings of a relationship?* Working Paper 498, School of Geography, University of Leeds.

Bogaerts, M.J.M. (1987),'Grootschalige ruimtelijke informatiesystemen in Nederland — Nieuwe ontwikkelingen in de periode 1984-1987' (Large-scale spatial information systems in the Netherlands — new developments in the period 1984-1987), *Kartografisch Tijdschrift,* 13, pp.38-42.

Briggs, D.J. (1988),'The use of broad-scale GIS for modelling the environmental impacts of land use change in Europe', Paper presented to the Polish Academy of Sciences, Radzikow, Poland.

Bunce, B, Vincent, P., Gatrell, T. and Dune, C. (1989), *GIS, forestry potential and the ITE land classes,* Research report Northern Regional Research Laboratory. (forthcoming).

Burrough, P.A. (1986), *Principles of Geographical Information Systems for land resource assessment,* Oxford University Press, Oxford.

Calkins, H.W. and Tomlinson, R.F. (1984), *Basic readings in Geographic Information Systems,* SPAD Systems, Williamsville.

Carver, S.J. (1988), *Siting issues in the disposal of low and intermediate level radioactive waste: the role of GIS,* Conference proceedings, ESRI User Conference 1988, Kranzberg, West Germany.

Charlton, M.E., Openshaw, S. and Carver S.J. (1989), 'Monte Carlo simulation of error propagation in GIS operations', Paper presented at the 29th European Congress of the Regional Science Association, Cambridge.

Clarke, K.C. (1986), 'Recent trends in geographic information system research', *Geo-processing,* 3, pp.1-15.

Clarke M. (1990), 'Geographic Information Systems and model-based analysis: towards effective decision support systems', in H.J. Scholten and J. Stillwell (eds), *Geographical Information Systems for urban and regional planning,* Kluwer Academic Publishers, Dordrecht.

Department of the Environment (1987), *Handling geographic information,* Report of the Committee of Enquiry chaired by Lord Chorley, HMSO, London.

Engebretsen, O. (1987), 'The Norwegian land accounting system', in *ECE-Seminar on new techniques to collect and process land-use data, Volume II: Response papers,* Gävle, Sweden.

Falk, T. (1987), 'Land-use information for physical planning and land consolidation in Sweden', in Proceedings of the Seventh Annual ESRI User Conference, Redlands, California.

Fedra, K. (1986), *Advanced decision-oriented software for the management of hazardous substances,* Part II: A demonstration prototype system, International Institute for Applied Systems Analysis, Laxenburg, Austria.

Fedra, K., Li, Z., Wang, Z. and Zhao, Z. (1987), *Expert systems for integrated development: a case study of the Shanxi,* SR-87-1, International Institute for Applied Systems Analysis, Laxenburg, Austria.

Gatrell, A.C. (1988), *Handling geographic information for health studies,* Research report no.15, Northern Regional Research Laboratory.

Green, N. (1988), 'Assessing Geographic Information Systems for the management and analysis of planning data', in Proceedings of the Eighth Annual ESRI User Conference, Palm Springs, California.

Grimshaw, D.J. (1988), 'The use of land and property information system', *International Journal of Geographic Information Systems*, 2, pp.58-65.

Groen, J. (1988), 'A strategy for investing in urban and rural development', in *Accessing the world,* Proceedings of GIS/LIS 1988, San Antonio, Texas, USA, pp.314-21.

Grossman, W. (1987), 'Dynamic geographical maps applied to forest die-off', in *Spatial Information Systems and their role for urban and regional research and planning*, Bundeskanzleramt, Vienna, pp.205-16.

Hitzert, M.C. (1989), 'CAD een goede zaak; ook voor gemeenten' (CAD a useful system, also for municipalities), *Automated mapping facilities management,* Proceedings of the First Dutch Conference on AM/FM, The Hague, pp.25-7.

Horn, D.R. and McLaren, G. (1988), 'Visual impact assessment in the commercial forest design and management process', in J.C. Muller (ed), *Environmental applications of digital mapping,* Proceedings of Eurocarto Seven, ITC, Enschede, The Netherlands, pp.160-71.

Koussoulakou, A. (1988), 'Simulation and dynamic mapping of air pollution for the area of Athens, Greece', in J.C. Muller (ed), *Environmental applications of digital mapping,* Proceedings of Eurocarto Seven, ITC, Enschede, The Netherlands, pp.184-93.

Marble, D.F. and Peuquet D.J. (1983), 'Geographic information systems and remote sensing', *Manual of Remote Sensing,* Society for Photogrammetry and Remote Sensing, USA.

Masser, I. and Blakemore, M.J. (1988), *The regional research laboratory initiative: the experience of the trial phase,* Economic and Social Research Council, UK.

Montanari, A. (1989), *Strategic project 'the Mediterranean: environment and economic development',* National Research Council, Naples.

Openshaw, S. (1987), *Analysing and exploiting client data by creating a geographic marketing information modelling system,* Economic and Social Research Council, UK.

Openshaw, S., Charlton, M. and Wymer, C. (1986), 'A geographical information and mapping system for the BBC Domesday optical disc', *Transactions of the Institute of British Geographers,* 11, pp.296-304.

Openshaw, S., Charlton, M., and Wymer, C. and Craft, A. (1987), 'A Mark 1 Geographical Analysis Machine for the automated analysis of point data sets', *International Journal of Geographical Information Systems* 4, pp.335-59.

Osinga, J.M. (1988), *Mogelijkheden voor het beoordelen van tracevarianten voor de hogesnelheidsspoorlijn met behulp van een geografisch informatiesysteem* (The potential for judging different routing alternatives of a high-speed train using a geographical information system), Wageningen Agricultural University, Wageningen, The Netherlands.

Ottens, H.F.L. (1988), 'A center for expertise for geographic information processing in The Netherlands', Paper presented at the Fifth International Workshop on Strategic Planning, ITC, Enschede, The Netherlands.

Ottoson, L. and Gut, T. (1986), 'Digital geographic information in Sweden', in B. Rystedt (ed), *Land-use information in Sweden. Applications of new technology in urban and regional planning and in the management of natural resources,* Swedish Council for Building Research, Stockholm, pp.45-52.

Padding, P. and Scholten, H.J. (1988), 'De toekomst van de landbouw in Ruimtelijk Perspectief' (The future of agriculture in a spatial perspective, with a summary in English), *Landschap,* 3, pp.201-11.

Parker, D.H. (1988), 'The unique qualities of a geographic information system: a commentary', *Photogrammetric Engineering and Remote Sensing,* 54, pp.1547-9.

Piscator, I. (1986), 'The Swedish land data bank system and its use by local authorities', in B. Rystedt (ed), *Land-use information in Sweden. Applications of new technology in urban and regional planning and in the management of natural resources,* Swedish Council for Building Research, Stockholm, pp.59-68.

Rhind, D. (1988), 'A GIS research agenda', *International Journal of Geographical Information Systems,* 2, pp.23-8.

Rhind, D. and Mounsey, H. (1989), 'GIS/LIS in Britain 1988', in P. Shand and R. Moore (eds), *The Association for Geographic Information yearbook 1989,* Taylor and Francis, London.

Schaller, J. (1986), 'Environmental impact of the proposed site for the 1992 Winter Games in Berchtesgaden', in Proceedings of the Sixth Annual ESRI User Conference, Redlands, California.

Schaller, J. (1987), 'Environmental Impact Assessment (EIA) of different routing alternatives of a planned autobahn between Munich and Passau (Southern Bavaria)', in Proceedings of the Seventh ESRI Annual User Conference, Redlands, California.

Schaller, J. and Haber, W. (1988), 'Ecological balancing of network structures and land use patterns for land-consolidation using GIS-technosystems: the North American experience', *International Journal of Geographical Information Systems,* 3, pp.203-18.

Scholten, H.J. (1987), 'Application of GIS in Regional Planning', in M.

Giaoutzi and P. Nijkamp (eds), *Development and applications of geographical information systems, informatics and regional development*, Gower Press, Aldershot, pp.292-305.

Scholten, H.J. and Meijer, E. (1988), 'From GIS to RIA', Paper presented at the URSA-NET Conference, Patras, Greece.

Scholten, H.J., and Padding, P., (1990) 'Working with GIS in a policy environment', *Environment and Planning B* (forthcoming).

Scholten, H.J. and van der Velde, R.J. (1987), 'Advanced applications of Geographic Information Systems in regional planning', in Proceedings of the Seventh Annual ESRI User Conference, Redlands, California.

Shaffer, C.A., Samet, H. and Nelson R.C. (1989), *QUILT: a geographic information system based on quadtrees*, University of Maryland, Maryland.

Southworth, F. and Chin, S-M. (1987), 'Network evacuation modelling for flooding as a result of dam failure', *Environment and Planning A, 19*, pp.1543-58.

TNO (1988), '*SEABEL: a hazard identification and decision support system for emergency response for chemical spills at sea,* TNO Division of Technology for Society, Apeldoorn, Netherlands.

Tomlinson, R.F. (1987), 'Current and potential uses of geographical information systems: the North American experience', *International Journal of Geographical Information Systems, 3*, pp.203-18.

Whitehead, C. (1988), *Population profiles around British Railway Network Stations,* Study Report no. 1, South-East Regional Research Laboratory, Birbeck College, London.

Chapter Three
GIS for urban and regional planning and analysis in Australia

Barry J. Garner

Introduction

Australia has been particularly responsive to the advances made during the past decade in the development of computer systems for handling geographical data. The application of land information systems (LIS) and geographical information systems (GIS) is now well established throughout Australia in both the public and private sectors. This is particularly the case in the area of LIS and in natural resource and environmental management but less so for social, economic and planning applications — these are areas in which developments are still at a relatively early stage compared to some other countries.

In this chapter, selected developments in LIS and GIS for urban and regional planning and analysis in Australia are reviewed. The emphasis is on GIS applications rather than on planning *per se*, the organisation of which varies considerably between the states and territories (see Neutze, 1978; McLoughlin and Huxley, 1986; Hamnett and Bunker, 1987). By and large, however, state governments control statutory planning and regional policy in each jurisdiction through their respective Departments of Planning (and Environment) and special purpose Agencies. Local governments are principally concerned with strategic planning and controlling development in compliance with the requirements of the State Acts. The Commonwealth government in Canberra has traditionally played an insignificant role in urban and regional planning although from time to time it establishes collaborative arrangements with the states usually in the context of special projects such as the New Cities programme of the 1970s.

GIS applications and planning

Geographical information systems have been defined in various ways by as many people. Generically they are computer-based systems for the capture, storage, manipulation, retrieval and display of spatial data (see Clarke, 1986; Cowan, 1988; Parker, 1988). A multiplicity of such systems has now been developed for a range of computing environments. Regardless of their specific technical details, GIS applications in planning may be located along a continuum in terms of their capabilities and the kinds of problems they address: this is examined in Figure 3.1.

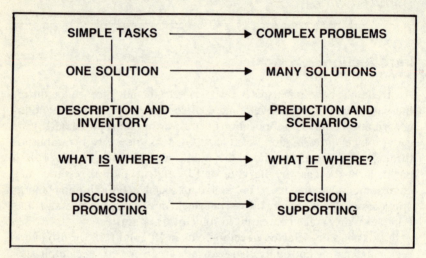

Figure 3.1 *An overview of GIS applications in planning*

Most LIS/GIS applications have been directed at tasks towards the left-hand side of Figure 3.1. This is also the area of the application of computer mapping. Emphasis is on answering questions of the 'what is' and 'where are' varieties which may simply involve the production of a thematic map of a single item. More commonly, however, questions encompass a combination of items, for example: Where are the places with a particular set of attributes? What are the characteristics of a particular place? How much of a particular attribute or combination of attributes is present in a particular location? Typically there is only one answer to these questions — essentially a description in map and record form of the existing situation. The capacity for GIS to address these kinds of questions in planning is now well documented, the answers to which are important for what may be thought of as the discussion-promoting role in decision making and policy formulation. At this level of application, GIS is used primarily for description and to provide users with inventory information.

Planners also need process information however, that is information describing change over time in attributes, their distribution and their interrelationships. This may be obtained from repeated inventories as part of the monitoring process or by analytical modelling to generate predictions and scenarios of future states in a given spatial system — this is a much more difficult task but one which is an integral part of the policy-making process. Given the present state of GIS technology, their use so far has been much more limited in addressing these more complex problems for which the central concern is one of invention rather than inventory. By the term 'invention', I mean the ability to answer questions of the 'what if' and 'where' variety: these are questions which are fundamental to spatial forecasting, evaluating alternative policies and the formulation of new plans. The concept of augmenting the cartographic modelling capabilities of GIS with the decision and choice capabilities explicit in mathematical modelling is novel but not new (Garner, 1982a). Although such integration is the subject of a growing body of research, the results have yet to be incorporated into commercially available systems. Because of this, it is argued that GIS has not yet realised its full potential as a decision-supporting tool in urban and regional planning analysis, and policy making.

In Australia, as elsewhere, the scope of GIS applications in urban and regional planning has been restricted as much by the capabilities of currently available systems as by institutional arrangements. Most applications have been to tasks at the left-hand side of the continuum in Figure 3.1; however, following research, particularly that by the Commonwealth Scientific and Industrial Research Organisation (CSIRO), a number of systems with limited mathematical modelling capabilities have recently been developed and successfully applied to urban and regional analysis. As a result, there has been a shift in applications towards the right hand side of Figure 3.1. Most of these systems are, however, best described as model-orientated GIS. It will be some time yet before truly model-based GIS are developed and their potential application in urban and regional planning and analysis is fully realised in practice. Indeed, the development of truly model-based GIS still constitutes the 'new frontier' for research and development in GIS technology (Tomlinson, 1984).

State-wide Land Information Systems

To date, the most active participants in the development of LIS/GIS have been in the public sector, particularly state government departments and agencies. In the Federal system of government in Australia, the six states and two territories have jurisdiction over all the major areas in which land and geographical information systems are typically applied: these areas include biophysical resources, social services and health, planning and economic development, transportation, utilities and property ownership. GIS is being applied progressively in all these areas if perhaps in an *ad hoc* manner. To date however, the major developments in GIS are not in Departments of Planning but in Lands Departments where the emphasis in all jurisdictions has been on the development of state-wide parcel-based LIS. Although these will most certainly be important for planning in the longer term, the immediate concern is the building of systems for fiscal administration and land taxation rather than for planning *per se*. Compared to many other countries, Australia has a sophisticated and tightly controlled system of recording land ownership based in large part on the Torrens system of land registration. Land parcel records are complete and comprehensively maintained in all states and territories, in part because the title to land is supported by government guarantees, but importantly because land and property taxes are a significant source of revenue for governments. The information contained in the cadastre is thus central to state land administration practices in Australia and basic to the requirements of local government and utility authorities (see Williamson, 1982).

Historically the responsibility for maintaining cadastral information has been shared between the various government agencies typically including the Land Titles Office, the Valuer General's Department, the Lands Department and, in the state capital cities, the Metropolitan Water Boards. Inevitably this has resulted in considerable duplication of records and responsibilities as well as inaccuracies between the records held by the different authorities in various manual administrative systems. The advantages of computer-based systems for the better management of cadastral information across departmental boundaries were quickly recognised.

Building on initiatives taken in the late 1970s, each of the six states and the two territories have progressively put in place the administrative and management structures for developing state-wide computer-based LIS during the 1980s. Typically, this has involved the creation of special bodies to oversee and coordinate activities. Examples of the creation of such coordination bodies are the State Land Information Council in New

South Wales which was established in 1985, and the Western Australian Land Information Secretariat which was created in 1981. Currently, each of the states and territories has operational cadastre-based systems although there are still wide variations in coverage, structure and implementation between jurisdictions, and these reflect differences in the political and institutional environments across the country. Historical trends and current developments in these systems are discussed in AURISA (1985), Williamson (1986), Williamson and Blackburn (1987) and in the URPIS volumes — the Proceedings of the annual Urban and Regional Planning Information Systems conferences (AURISA, 1976ff).

The development of state-wide LIS is the only area in which Australia can be considered to have a national programme in the application of computers to spatial data handling. Since 1986, the Australian Land Information Council (ALIC) has provided the formal mechanism to achieve coordination in LIS development and land management activities between the states. In 1987, ALIC endorsed a set of policies and procedures which now form the National Strategy for Land Information Management (ALIC, 1987). The principal aims of the National Strategy are to encourage cost-efficient access to land information throughout Australia; to provide an operational basis for effective decision-making on the social and economic use of land at all levels of government (and by the private sector); to develop the appropriate technology; and to provide the institutional mechanisms for efficient data transfer between users. Ultimately it is envisaged that the National Strategy will be enlarged to encompass a broader range of data bases than just the cadastre, hence additional government agencies including planning departments, and that applications will be possible at different geographical scales. The National Strategy is thus considered to be an evolving one.

LIS concepts

Although differing in detail, the LIS being developed across Australia generally conform to the model shown in Figure 3.2. Conceptually they comprise two sets of data (textual and graphic), the most important components in which are the 'legal' land parcels embodied in the digital cadastral data base (DCDB) and the various attributes attached to these. These are linked by a set of two-way pointers — the parcel number or property identifier in the case of land parcels and some form of geocode for other data items. Both are typically stored in each data set so that operationally the separation of textual and graphic information is transparent to the user (see Zwart, 1986; Zwart and Williamson, 1988).

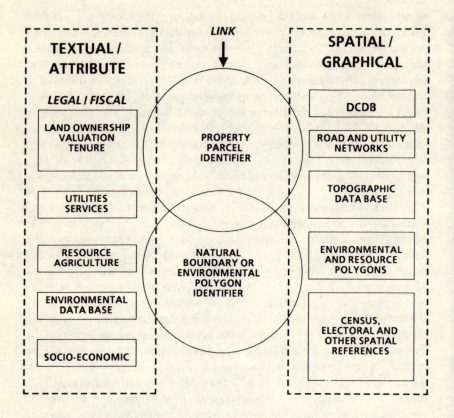

Source: I.P. Williamson and J.W. Blackburn 1987

Figure 3.2 *The LIS concept in Australia*

The key physical component in the systems being developed in Australia is the concept of the LIS Hub which has been designed to facilitate the transfer of land-related data between government departments and other authorities as well as providing the mechanism for efficient data administration and LIS management. Since the data in a LIS comes from many different sources, a key feature in the Hub concept is the identification of 'data trustees' which are the designated agencies responsible for providing particular items of data to the system, updating that data and establishing data standards. In this way data can be synchronised from different sources for transfer to user organisations.

The significance of the LIS initiatives taken by state governments for planning are obvious — at least in the longer term. When fully operational, these systems will provide state-wide coverage of up-to-date and accurate information at the legally defined land-parcel level. In the New South Wales LIS, for example, this will eventually encompass about three million parcels covering all 175 local authority areas within the state. Attached to each parcel, the area and location of which are given in the DCDB, will be information on ownership, valuation, land-use and development controls, details of buildings and structures, sales data and other planning-relevant data. In the New South Wales LIS, currently some 30 different attributes are attached to each land parcel in the system.

Given the current emphasis on the development of LIS for managing land records and fiscal administration, most systems have not yet been extended to include socio-economic data, natural resource data and environmental data. To overcome this deficiency some states have started to build separate systems specifically for urban planning. A good example is the industrial and commercial land information systems developed for the Perth metropolitan area by the Western Australia Department of Planning and Urban Development which is used for strategic planning (Rogers, 1989a). The overall objectives of the State Planning Commission's Industrial and Commercial LIS are to provide an accurate assessment of the amount of land either used for or zoned for various industrial and commercial uses; to provide a system which can be used to store, to access and continually to update land information; and to establish a data base which can be used to 'assist in the review of strategies and policies for the optimal location and development of industrial and commercial activities' (Rogers, 1989b).

The Perth system was developed in two phases. First, a realisation by the State Planning Commission that it lacked reliable industrial land information caused the Commission to establish an annual survey of industrial land from 1985 (this system contains about thirty variables on some 15,000 sites located in 74 industrial complexes). In 1987, it was decided to extend the data base to include all commercial activities and so a survey of some 21,500 commercial establishments was undertaken (though this survey excluded the establishments in the Perth Central Business District). Rogers (1989b) listed the advantages of the Western Australia system as achieving greater consistency and accuracy than earlier systems, reducing data collection and data maintenance costs, providing a flexible base for different user types and facilitating the future integration of other application areas and data sets. In addition to locally derived information sources, the five-yearly Census of Housing and

Population has been available in digital form since 1981 for ten categories of spatial unit in the Australian Standard Geographical Classification. This can be easily incorporated with parcel information and other data bases to provide an integrated geographic data base for urban planning applications.

Local government planning

Although the emphasis on parcel-based LIS in Australia is at present of limited value in regional planning for which aggregated data at smaller scales is more appropriate, these micro-scale systems are particularly significant for planning in local government. The high resolution of parcel data is particularly well suited to the administration of building applications and development control as well as to the preparation of local strategic plans — these are the two principal planning activities. However, although the use of computers by local government has increased dramatically during the past decade, their use in planning is generally at an early stage of development.

The results of a 1987 survey covering 718 of the 916 legally defined Local Government Authorities in Australia indicated, for example, that whereas at least 84 per cent of these were using computers (a threefold increase since 1977) only 32 per cent of local government authorities used computers (mainframe or micros) for some planning-related activity (Murphy *et al.,* 1988; Earle *et al.,* 1986). The use of computers by local government is still dominated by administrative transaction processing applications particularly in the area of financial management. Function-specific applications are still rather limited and, in planning, the use of information systems is oriented more to development control than to strategic planning.

The application of computers to spatial data analysis (e.g. small area population and dwelling forecasts or demographic modelling and demand analysis for public facility provision) and mapping is still relatively underdeveloped, while the operational use of modelling techniques and GIS is still exploratory and the exception rather than the rule. Many small-scale pilot projects are, however, now under way. A major aim of these has been to demonstrate the ways in which land-parcel data can be aggregated, integrated with census and other kinds of socio-economic data, and processed for local planning applications using GIS data manipulation and analysis capabilities. Examples of this include demonstration projects in Perth (Devereux, 1985), Hobart (Zwart and Williamson, 1988), the City of Brighton, Melbourne (Newton *et al.,*

1988) and Coopers Plains, a suburb of Brisbane (Perrett *et al.*, 1989). Mirroring the activities by the state governments the emphasis at the local government level is still primarily focused on the development of general purpose LIS typically configured on the model shown in Figure 3.3.

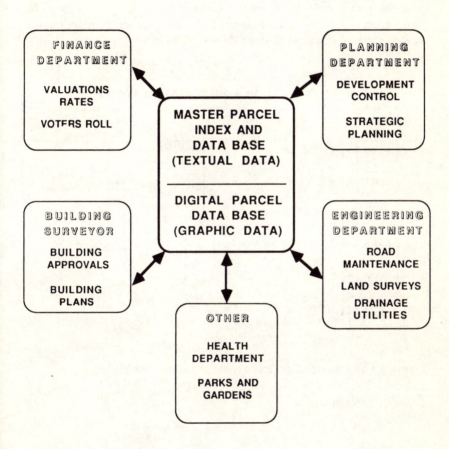

Source: P.R. Zwart and I.P. Williamson, 1988

Figure 3.3 *The structure of general purpose LIS for local government use*

The Sydney City Council LIS

The development of LIS is a priority activity of local government located in the heart of the capital city metropolitan regions (see Hanna and Wagner, 1985, on Adelaide; Jankovits, 1986, on Brisbane; and Butler, 1989, also on Brisbane). These systems are essentially modelled on the earliest — and arguably still the best — example of the application of spatial information systems in local planning in Australia, that developed over the previous 16 years by the Council of the City of Sydney. Its jurisdiction covers an area of about six square kilometres including the Sydney Central Business District (Nash, 1986).

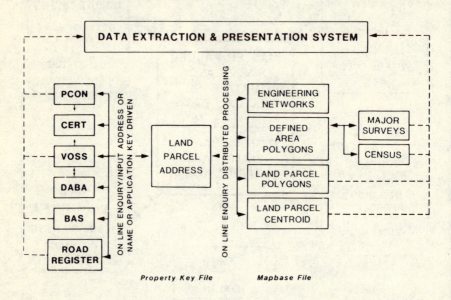

Figure 3.4 *The Sydney City Council planning information system*

Source: K. Nash, 1986

The planning functions of the system are outlined in Figure 3.4. Land parcels digitised at 1:100,000, and their associated property key, form the map base of the system. The attributes attached to these land parcels are accessed on-line to service a range of planning functions. Subsystems for use in development control and building application include the Valuation, Ownership and Sales System (VOSS), a non-spatial property information system providing details of six key items for each land parcel which is also used automatically to produce Certificates of Compliance

(CERT) in accordance with the provisions of the New South Wales Environmental and Planning Act. VOSS is linked to the Development Application/Building Application System (DABA) which is used to monitor these requests and to streamline procedures in the Planning Department and to the Planning Control System (PCON) which contains details of all statutory and non-statutory planning controls (there are currently over 160 planning instruments affecting land under the Council's jurisdiction) for each parcel since 1971.

The subsystems which serve the planning and research needs encompassed by the City of Sydney Strategic Plan include the Building Activity System (BAS) — which provides a detailed description of land-use, building fabric and employment for the more than 60,000 establishments in the City — and the Data Extraction and Presentation System (DEPS). This interactive, spatially-orientated general enquiry system is designed to extract data from the other subsystems and from external data sources (such as the census) for analysis and display in the planning context. The capacity for spatial information processing in DEPS is, however, rather limited given the lack of a true GIS capability. Nevertheless, the Council's system has proved to be an invaluable asset for its planning and engineering activities and it provides a good example of the capabilities of spatially-orientated information systems for other local government organisations to emulate.

Examples like the Sydney LIS, the Brisbane LIS and the Western Australia Commercial and Industrial LIS show that LIS development in Australia is quite well advanced and that many local and state government bodies have invested heavily in the creation of such systems. Butler (1989) has pointed out that since the Brisbane LIS has been developed and progressively implemented, the demand for accurate and up-to-date information has grown substantially and that this has culminated in the significant extension of the original computer network throughout Brisbane. In Brisbane, as in other parts of Australia, the LIS hub approach will link most sources of data enabling access to land data (initially) and other sources of spatial data (eventually) by those planners and decision-makers who require a continuous flow of information to help them in their tasks. Indeed, Brisbane City Council's goal for the implementation of the Brisbane LIS is that the 'Council will ensure the potential of the system is fully realised, not only for map production, but also for on-line enquiries, data analysis and urban management generally, which will result in cost efficient government to the benefit of all' (Butler, 1989).

Urban and regional analysis

The development of planning models is well established in Australia although their application is not as well represented in the activities of state planning departments as might be hoped or expected. In this respect, Australia is typical of many other countries (see Breheny, 1987). The main focus in the use of models for urban and regional analysis is the research undertaken in the Universities and especially in the various divisions of the Commonwealth Scientific and Industrial Research Organisation (CSIRO). These organisations have been engaged for many years in research and development in model building, geographical information systems, and the broader aspects of spatial data handling.

The CSIRO Division of Building, Construction and Engineering in Melbourne, has had a long history of research into model-building and software development for spatial data handling, particularly in modelling urban land-use, transport, infrastructure, planning and for mapping. Representative contributions include TOPAZ, an urban activity location model (Brotchie *et al.*, 1980); LAIRD, a package for assessing impacts of retail development (Roy and Anderson, 1988); MULATM, a traffic planning package which is highly reliant on interactive graphic displays at all levels of operation (Taylor, 1988, 1989); and POLDIF, a model for conducting environmental impact analysis (Taylor and Anderson, 1988). Many of these have been converted for use on microcomputers, and in the future their application by planning departments is expected to increase as 'Desktop Planning' becomes more widespread (see Newton *et al.*, 1988). However, most of these and other models have not so far been incorporated into GIS but operate as stand-alone systems. Most of these models also have a relatively primitive capability for graphic output.

Transportation planning is an area in which the integration of GIS and model-based analysis is proving particularly successful. The 'Transport Planning Microscope' (TPM), which is described in Taylor (1989), is a PC-based system designed for interactive use which includes both GIS software and a connected hierarchy of traffic and transportation modelling packages which enables the planner/analyst to shift the focus of investigation between the micro and macro levels of analysis. TPM comprises two transport network models (MONTRANP and MULATM) and an interactive transport data analysis and display package (DIAMONDS — which according to Taylor (1989) is 'an interpretive tool for use with a combined physical, demographic and transport database constructed in a GIS framework'). The system has links to other models; these include TRANSTAT (for the statistical analysis of transport data), POLDIF (for estimating the environmental impact of road traffic), MULTSIM (for the analysis of arterial road flows) and SIDRA (for road intersection analysis and design).

Indeed, TPM provides a 'powerful means for the capture, display, analysis, retrieval, modelling and interpretation of data relating to the way people use transport systems, and the performance of these systems' (Taylor, 1989). In addition to its role as a means of integrating various software packages, the TPM system offers new opportunities for integrated data analysis using a variety of data sources which range from topographic data systems, through land and property data systems, to transport network information, to social and demographic data, to transport flow data and to environmental data. Taylor (1989) pointed out that the GIS element of the TPM system provides a framework for integrating several transport-related data bases and for supporting various modelling activities ranging from land-use allocation through to environmental impact analysis.

One area in which there has been significant attempts to incorporate models into a GIS framework is in assessing the planning implications of natural hazards. Many Australian communities are prone to floods and bushfires, the effects of which are important considerations in land-use planning at the sub-regional scale. The cost of bushfires — in both human and economic terms — is particularly high throughout the country, especially in suburban areas and country towns. Not surprisingly, a considerable research effort over the recent past has been made to develop models for predicting and monitoring bushfire behaviour. One of these, PREPLAN (Kessell, 1988), has also proved to be quite useful in determining the environmental sensitivity of potential areas for residential development in the urban-rural fringe. Using the cartographic modelling capabilities of GIS, PREPLAN combines data on fire risk with those for potential urban development to identify areas where stringent development controls must be enforced. Similar approaches using GIS have been used to identify areas at risk from floods in urban communities (Smith and Greenaway, 1988).

Another good example of the application of GIS that bridges the gap between analysis and implementation in the planning context is the LUPIS package, also developed by the CSIRO (Ive and Cocks, 1988). LUPIS is a decision support system for land-use planners and managers which allocates preferred land-uses to predefined mapping units in an area depending on the importance attached to alternative policy criteria. LUPIS is increasingly being used by planners in Australia for the allocation and management of land resources. Although to date most applications have been for environmental planning, the system can be readily extended to urban land-use problems. Its main strength is in the ability to incorporate decision-making rules into the traditional GIS cartographic modelling capabilities. A more advanced application has been developed separately for producing zoning schemes based on the knowledge-based systems widely used for organisational management purposes (Davis and Grant, 1987).

Territory assignment

The division of space into smaller units or territories has been an important prerequisite for many planning activities, for example, the production of zoning maps for strategic planning and the delimitation of larger territories for urban and regional planning. In Australia, an 'official' set of regions has been established for planning purposes in all the states and territories. Typically these have been delimited in an *ad hoc* way based on the two largest general purpose geographic units in the Australian Standard Geographical Classification — the statistical subdivisions and statistical divisions used by the Australian Bureau of Statistics for presenting the results of the census (see Brauer, 1985). Although planning regions based on these have proved to be useful for general planning purposes, from time to time special purpose regions are required for particular tasks.

Region building has had a long history in planning and has been the subject of considerable research over the years, particularly by geographers. Early attempts to delimit planning regions were typically based on the analysis of multivariate data using techniques like factor analysis and principal components analysis, the results from which were input to a taxonomy programme to identify spatially contiguous clusters which were then mapped as regions. A good example of this approach in Australia is the delimitation of regions by the Department of Urban and Regional Development which was established after the Australian Labour Party won the 1972 national elections. The regions were used as the basis for allocating general assistance grants to local authorities (see Logan *et al.*, 1975).

In these early attempts at region building, the computer was used solely for data analysis and the techniques used did not always generate an optional set of regional boundaries. These earlier approaches have now been superseded by and large by the application of GIS to the general problem of delimiting regions in a variety of contexts — a task widely referred to as territory assignment. There are several examples of the development and application of special purpose GIS for this problem in Australia including the Australian Resources Information System

(ARIS) and the Interactive Territory Assignment System (ITA) — both developed by researchers in the CSIRO.

The Australian Resources Information System (ARIS)

ARIS is a continental scale computerised GIS developed over the past decade which is designed to answer questions about the location within very large parts (or all) of Australia of specified combinations of biophysical and socio-economic resources using either polygon or grid cell map bases. The system currently includes ten separate data bases including the Australian Resources Databank containing over 400 biophysical data items embracing terrain characteristics such as elevation, soil type, lithology, land cover and vegetation types and level of soil degradation. This data base is also linked to socio-economic data from the census. ARIS has been used to support a variety of studies by the Commonwealth government including mapping the effects of the location and orientation of the AUSSAT communication satellite, rangeland management, the identification of sites for new cities and, more recently, for the delimitation of urban regions (Cocks and Walker, 1987, 1989).

The latter application formed part of the Commonwealth Government's Country Centres Project, the broad purpose of which was to establish a data base for regional monitoring across Australia, to provide information to local government bodies and regional authorities to assist them in their various planning and management tasks, as well as to alert the Commonwealth Government to regional economic problems and to assist the Commonwealth Government in assessing regional development potential. A necessary prerequisite for this process to occur, however, was the delimitation of an appropriate set of mutually exclusive regions appropriate for urban economic analysis at the national level. Seventy-five towns, cities and metropolitan areas were selected as nodes for these regions, the spatial building blocks for which were local government areas. Using the spatial searching routines in ARIS, centroids of local government areas were allocated to the nearest urban node to identify iteratively the set of regions. The boundaries of the resultant regions were then modified as necessary to take account of the effects of natural barriers and major communication networks (Parvey *et al.*, 1988). The resulting territory assignment is shown in Figure 3.5. Subsequently, ARIS was used to develop natural resource and socio-economic profiles for each region to assist the Commonwealth Government in the formulation of national economic development policy.

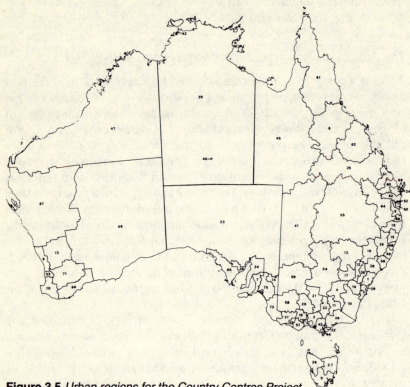

Figure 3.5 *Urban regions for the Country Centres Project*

Source: Parvey *et al.*, 1988

The Interactive Territory Assignment System (ITA)

Whereas the application of ARIS for territory assignment was based principally on standard GIS capabilities, ITA was developed specifically to incorporate a modelling capability in GIS that would permit the identification of a set of optimal regions using location-allocation techniques. ITA is consequently a model-oriented GIS developed specifically to address a wide range of territory assignment problems. It has now been applied successfully in reorganising the spatial structure of major banks, to delimit the boundaries of electoral regions and to identify a set of optimal regions for the reorganisation of the administrative territories used for managing the provision of fire services by the New South Wales Department of Fire Brigades (Horn *et al.*, 1988).

This latter application was undertaken as part of a management review of the fire department. The objective was to identify whether the number of regions used to administer the provision of fire brigades in New South Wales (currently 10 regions) could be reduced and, if so, to identify what number of regions there should be in the future and how these regions should be delimited geographically. The constraint was that the workloads of the Superintendents responsible for maintaining the level of fire service provision in each region should be approximately equal and not to exceed 42 hours per week — including both time spent on administrative tasks and that spent travelling within the region to carry out delegated responsibilities. The potential headquarters (major country towns) for the regions were predetermined, hence the problem was essentially one of assigning local government areas to selected towns to create a set of territories satisfying the constraints on workload. By making use of the Teitz-Bart algorithm to solve the p-median problem in location-allocation modelling, ITA established that nine regions were the minimum to satisfy the objective function of minimising travel times within each region and ensuring equality of workloads between them. The initial regionalisation and the optimal set of regions are shown in Figure 3.6.

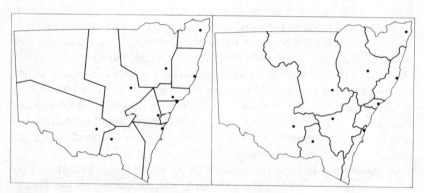

Figure 3.6 *Initial (left) and optimal (right) regions for the provision of fire services in New South Wales*

Towards model-based GIS

The previous two examples demonstrate how conventional GIS are being used in urban and regional analysis in Australia and the way in which the traditional cartographic modelling capabilities of GIS are being augmented by the addition of mathematical modelling capabilities to expand their application to new problem areas. The development of these 'second generation' GIS opens up a number of potentially new areas of application in urban and regional planning and analysis. In terms of the schema shown in Figure 3.1, these model-oriented GIS are shifting applications towards the right-hand side of the diagram and towards addressing inherently more complex problems for which there may not be a single answer. However, a different kind of modelling capability is required if GIS are to be used successfully for prediction and scenario development in the planning process — the development of which is still at a very early stage in Australia and overseas.

An illustration of the potential application of model-oriented GIS is for the monitoring of the regional development process for planning purposes and, for example, that resulting from the expansion of coal mining in the Hunter Valley, New South Wales. This region, located to the north of Sydney, contains the largest black coal reserves in New South Wales. In 1980, firm commitments had been announced to develop around twenty new, large open-cut coal mines in the region by 1985. The mines were expected to employ an additional 3,500-4,000 workers directly, in addition to the jobs created indirectly from multiplier effects. As it turned out, the boom in the coal industry did not materialise and many of the anticipated projects did not proceed. However, enough of them did proceed to cause severe strains on the region's infrastructure and to present a number of major problems in planning to accommodate the resultant population growth. The need to be able to evaluate and monitor the cumulative impacts of coal mine development on the physical and human environment in the region were clearly essential for proper and effective regional planning.

A conceptual framework suitable for evaluating and monitoring the socio-economic effects of the expansion of coal mining in the Hunter Valley is shown in Figure 3.7 (Garner *et al.*, 1981). Starting at the top with the existing regional environment, the introduction of a new coal mine sets in train a series of linked effects, which, when played out in the region, result in an altered regional environment. This then becomes the starting-point for the next coal mine to be developed. The problem, of course, is that the development process is not as rigidly sequential as the model would imply, as many new coal mines are developed

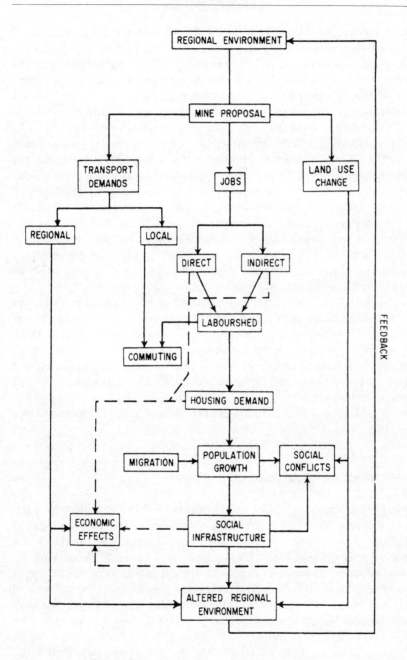

Figure 3.7 *A general model for monitoring the impact of new coal mines in the regional development process*

simultaneously and as a result the feedback process is quite complex. A critical expression of this is the alteration to the housing stock and the regional housing market which both condition the residential distribution of the work-force. This then affects journey-to-work which, ultimately, affects the spatial pattern of population growth and the associated demand for social and physical infrastructure in the region.

The modelling of this kind of process requires a different type of GIS to those currently available, one in which various sub-models are included — for example entropy-maximising models for allocating people to different settlements in the region — as well as others which together can effectively simulate the cumulative nature of the regional development process using different assumptions. The application of simulation models in planning would appear to lag behind that in the environmental sciences. The development of the kind of model-based GIS with in-built spatial forecasting, location-allocation and other optimising capabilities is clearly of fundamental significance for future applications of GIS in regional planning and analysis. Such GIS would permit a more effective monitoring of developments enabling the speedy revision of estimates and predictions resulting from cut-backs in the scale of projects and set-backs in the timing of developments. More important, the kind of GIS envisaged here would enable planners and policy makers to invent alternative scenarios of the regional development process by simulating these using different assumptions and by asking alternative 'what if' questions (Garner, 1982b). This mode of analysis and the development of 'intelligent' GIS is discussed in detail by Birkin et al. (elsewhere in this volume).

Retrospect and prospect

Apart from the significant developments of LIS by state and local governments, compared to many other countries the widespread application of GIS in urban and regional planning is still perhaps at a relatively early stage of development in Australia. However, an infrastructure has been put in place and there is a growing awareness of the significance of GIS in state and local governments for urban and regional planning and analysis. The availability of microcomputers will undoubtedly hasten the diffusion of 'Desktop Planning' and widen the application of GIS technology.

The release in 1988 of CDATA86 by the Australian Bureau of Statistics has already had a significant impact in extending the use of computers in planning applications. This package comprises a range of items from the 1986 census plus selected 1981 data on CD-ROM together with a colour mapping package which enables a wide range of statistical profiles and maps to be produced for a variety of different spatial units including postcode areas. Australia still lacks, however, a good address matching capability. There is no equivalent to the DIME and TIGER files used by the US Bureau of Census for example — although a number of private sector organisations are currently developing the equivalent to these. When available, these will undoubtedly expand the application of GIS into entirely new areas of planning-related activities.

Research and development in GIS is now well advanced, especially in the development of model-oriented GIS. Although this will undoubtedly result in a widening of applications in urban and regional planning and analysis, these will be restricted in scope given the predominantly area-oriented (polygon and raster) nature of current GIS software. The full realisation of GIS for planning applications awaits the development of point-based systems, which will permit their extension to problems focusing more directly on planning for people rather than places, as well as the development of truly model-based GIS for predictive purposes. But these are limitations that transgress national boundaries and are not peculiar to Australia.

Despite its relative isolation in physical terms, Australia is keeping abreast of international trends in GIS developments and applications, and is poised to embark on a major expansion in the use of GIS in the coming decade in all application areas, both public and private, the effects of which undoubtedly will be felt particularly in the field of urban and regional planning and analysis.

Bibliography

ALIC (1987), *National strategy for land information management,* Australian Land Information Council Secretariat, Canberra.

AURISA (1976 ff), *URPIS 1-17,* Australasian Urban and Regional Information Systems Association Inc., Sydney.

AURISA (1985), *Report of the working group on statewide parcel-based land information systems in Australasia,* Australasian Urban and Regional Information Systems Inc., Sydney.

Brauer, A. (1985), 'Introduction of the Australian standard geographical classification (ASGC)', *URPIS 13,* AURISA, Sydney, pp.365-97.

Breheny M.J. (1987), 'The context for methods: the constraints of the policy process on the future of quantitative methods', *Environment and Planning A,* 19, pp.1449-62.

Brotchie, J.F., Dickey, J.W. and Sharpe, R. (1980), *TOPAZ — a general planning model and its applications at the urban and facility planning levels,* Heidelberg, Springer-Verlag.

Butler, J.L.T. (1989), 'Brisbane City Council digital mapping system: a major tool for urban planning and management in Brisbane', Paper presented at the International Conference on Computers in Urban Planning and Urban Management, Hong Kong.

Clarke, K.C. (1986), 'Advances in geographic information systems', *Computers, Environment and Urban Systems,* 10(3/4), pp.175-84.

Cocks, K.D. and Walker, P.A. (1987), 'Using the Australian Resources Information System to describe extensive regions', *Applied Geography,* 7(1), pp.17-27.

Cocks, K.D. and Walker, P.A. (1989), 'ARIS: a working geographic information system for continental Australia', in D. Ball and R. Babbage (eds), *Geographic information systems: defence applications,* Pergamon Press, Sydney, pp.129-51.

Cowan, D.J. (1988), 'GIS versus CAD versus DBMS: what are the differences?', *Photogrammetric Engineering and Remote Sensing,* 54(11), pp.1551-55.

Davis, M.R. and Grant, I.W. (1987), 'ADAPT: a knowledge-based decision support system for producing zoning schemes', *Environment and Planning B,* 14(1), pp.53-66.

Devereux, D. (1985), 'The urban two project', *URPIS 13,* AURISA, Sydney, pp.225-34.

Earle, T.R., Fitzgerald, E.P. and Learmonth, R.D. (1986) 'Who is using computers in local government?' in P.W. Newton and M.A.P. Taylor (eds), *Microcomputers for local government planning and management,* Hargreen, Melbourne, pp.14-21.

Garner, B.J. (1982a), 'Towards geographic information systems for evaluating the socio-economic effects of resource developments', *Land Use Modelling Quarterly,* 4(3), pp.74-80.

Garner, B.J., (1982b) 'Towards more people-oriented geographic information systems', *URPIS 10,* AURISA, Sydney, pp.200-11.

Garner, B.J., Holsman, A., Phibbs, P. and van Kempen, T. (1981), 'Assessing the socio-economic impacts of coal mine developments with special reference to the Upper Hunter NSW', in J. Hannan (ed), *Environmental controls for coal mining,* Australian Coal Association, Sydney, pp.290-311.

Hamnett, C. and Bunker, R. (1987), *Urban Australia: planning issues and policies,* Mansell, London.

Hanna, E.J. and Wagner, G.M. (1985), 'The development of a land-based information system for the City of Adelaide', *URPIS 13,* AURISA, Sydney, pp.91-109.

Horn, M., O'Callaghan, J. and Garner, B.J. (1988), 'Design of integrated systems for spatial planning tasks', in D.F. Marble (ed), *Proceedings,* 3rd International Symposium on Spatial Data Handling, International Geographical Union, Columbus, USA, pp.107-16.

Ive, J.R. and Cocks, K.D. (1988), 'LUPIS: a decision-support system for land planners and managers', in P.W. Newton, M.A.P. Taylor and R. Sharpe, (eds), *Desktop planning,* Hargreen, Melbourne, pp.129-39.

Jankovits, E. (1986), 'Town plan preparation using digitised cadastral data', *URPIS 14,* AURISA, Sydney, pp.376-9.

Kessell, S.R. (1988), 'Fire hazard modelling: the PREPLAN and FIREPLAN systems', in P.W. Newton, M.A.P. Taylor and R. Sharpe (eds), *Desktop planning,* Hargreen, Melbourne, pp.222-31.

Logan, M.I., Maher, C.A., McKay, J. and Humphreys, J.S. (1975), *Urban and regional Australia: analysis and policy issues,* Sorrett, Melbourne.

McLoughlin, J.B. and Huxley, M. (1986), *Urban planning in Australia: critical readings,* Longman Cheshire, Melbourne.

Murphy, P.A., Zehner, R.B., Robinson, P.A. and Hirst, R. (1988), *Computer use by local government planners: an Australian perspective,* School of Town Planning, University of New South Wales, Kensington.

Nash, K. (1986), 'The application of computers to the planning tasks in the City of Sydney', *Australian Planner,* 24(3).

Neutze, M. (1978), *Australian urban policy,* George Allen and Unwin, Sydney.

Newton, P.W., Davis, J.R., Simberg, K.V. and Crawford, J.R. (1988),'A microcomputer-based integrated land information system', in P.W. Newton, M.A.P. Taylor, and R. Sharpe (eds), *Desktop planning,* Hargreen, Melbourne, pp.54-64.

Newton, P.W., Taylor, M.A.P. and Sharpe, R. (1988), *Desktop planning: microcomputer applications for infrastructure and services planning and management,* Hargreen, Melbourne.

Parker, H.D. (1988), 'The unique qualities of a geographic information system: a commentary', *Photogrammetric Engineering and Remote Sensing,* 54(13), pp.1547-9.

Parvey, C.A., Walker, P.A. and Pietrzak, B.E. (1988), 'The Country Centres Project: biophysical aspects of a national data base for assessing regional comparative advantage', *URPIS 15,* AURISA, Sydney, 14, pp.1-29.

Perrett, P., Lyons, K.J. and Moss, O.F. (1989), 'Overview of LIS activities in Queensland', in D. Ball, D. and R. Babbage (eds),

Geographic information systems: defence applications, Pergamon Press, Sydney, pp.152-79.

Rogers, M.B. (1989a), 'The impact of land information systems in land use planning in Western Australia', *URPIS 17,* AURISA, Sydney.

Rogers M.B. (1989b), 'The application of land information systems in planning in Western Australia', Paper presented at the International Conference on Computers in Urban Planning and Urban Management, Hong Kong.

Roy, J.R. and Anderson, M. (1988), 'Assessing impacts of retail development and redevelopment', in P.W. Newton, M.A.P. Taylor and R. Sharpe (eds), *Desktop planning,* Hargreen, Melbourne, pp.172-9.

Smith, D.I. and Greenaway, M.A. (1988), 'The computer assessment of urban flood damage: ANUFLOOD', in P.W. Newton, M.A.P. Taylor and R. Sharpe (eds), *Desktop planning,* Hargreen, Melbourne, pp.239-50.

Taylor, M.A.P. (1988), 'Computer models for traffic systems applications', in P.W. Newton, M.A.P. Taylor and R. Sharpe (eds), *Desktop planning,* Hargreen, Melbourne, pp.264-98.

Taylor, M.A.P. (1989), 'Traffic planning by a "desktop expert"', Paper presented at the International Conference on Computers in Urban Planning and Urban Management, Hong Kong.

Taylor M.A.P. and Anderson M. (1988), 'Modelling the environmental impacts of road traffic with MULATM-POLDIF: a PC-based system with interactive graphics', *Environment and Planning B* 15(4), pp.413-31.

Tomlinson, R.F. (1984), 'Geographic information systems — a new frontier', *The Operational Geographer,* 5(1), pp.31-6.

Williamson, I.P. (1982), 'The role of the cadastre in a statewide land information system', *URPIS 10,* pp.285-305.

Williamson, I.P. (1986), 'Trends in land and information system administration in Australia', *Proceedings,* Auto-Carto London, pp.71-82.

Williamson, I.P. and Blackburn, J.W. (1987), 'Current trends in developments in land information systems in Australia', *Proceedings,* 21st Conference, Institute of Australian Geographers, pp.289-97.

Zwart, P.R. (1986), 'Parcel-based land information systems', in P.W. Newton and M.A.P. Taylor (eds), *Microcomputers for local government planning and management,* Hargreen, Melbourne.

Zwart, P.R. and Williamson, I.P. (1988), 'Parcel-based land information systems in planning', in P.W. Newton, M.A.P. Taylor and R. Sharpe (eds), *Desktop planning,* Hargreen, Melbourne, pp.44-53.

Chapter Four
The Vermont GIS: a model for using Regional Planning Commissions to deliver GIS in support of growth management
Thomas L. Millette

Introduction

The Vermont State Geographic Information System (Vermont GIS) is the first state-wide geographic information system (GIS) to be legislatively mandated and funded in the United States. Although a number of states have GIS programmes (e.g. Illinois, New Hampshire, New Jersey, North Carolina, Texas, Utah, Washington, Maryland, Minnesota, Wisconsin, Massachusetts, Connecticut, Michigan, Rhode Island), these are fundamentally different from the Vermont GIS in that they usually consist of a single central service bureau which is generally nested within a state planning or natural resource agency where the GIS applications are restricted to an extremely narrow focus.

For example, the Connecticut Geographic Information System is a typical example of the type of state GIS programme currently found in the United Sates. The Connecticut GIS is located in the Natural Resources Division of the Connecticut Department of Environmental Protection. It has no legislative mandate, and its applications are primarily limited to stream flow modelling and water supply planning. Because the GIS has no official source of funding and because the operational budget is drawn from various sources within the Connecticut Department of Environmental Protection, funding is constantly in jeopardy. Furthermore, inadequate funding since the programme began in 1986 has resulted in the overall development of the system taking longer than expected due to delays in data base development, in the acquisition of adequate hardware and in obtaining software training (Croswell and Archer, 1988, pp.2-4).

The Vermont GIS programme differs from programmes in other states in that it has both a legislative mandate and an official source of funding. Additionally, the development of the GIS is being guided by a comprehensive plan which addresses the technical, political and administrative concerns which arise from the implementation of a programme which is intended to serve, and be subsidised by, activities at the state, regional and municipal levels. Most important, the Vermont GIS has a foundation which is built upon the recognition that a GIS is an essential planning tool. There is also a recognition that a strong planning input is essential to effective growth management and unless growth is positively managed, the state may lose the environmental and cultural attributes which contribute to Vermont's uniqueness. These recognitions account to a large degree for Vermont's ability to consolidate support at the state, regional and local levels for a cooperative programme to integrate GIS into planning activities for effective growth management.

Growth management: a context for the Vermont GIS

The amount of effort, cooperation and coordination on the part of state, regional and local officials, planners and citizens required to develop, implement and support the Vermont GIS has been extraordinary. The motivation of the large number of people and agencies needed to support the Vermont GIS required a special catalyst. This catalyst was a concern for the preservation of the Vermont way of life in the face of pressure for rapid growth and development. The history of recent growth management activities in Vermont is an important backdrop to any discussion of the development of the Vermont GIS, since it is from these activities that the formal GIS programme evolved.

Governor's Commission on Growth

On 22 September 1987 the Governor of Vermont signed an Executive Order that created a Commission to establish guidelines for Vermont's future growth and development. The twelve-member commission was charged with the task of holding state-wide public hearings and submitting a report to the Governor which would undertake three main tasks. The first task was to produce an overview of the current growth patterns in Vermont and to assess the effectiveness of existing laws and practices in managing that growth. The second task was to produce a statement of the goals and principles required to preserve Vermont's character. Among the specific issues the Commission was to consider were: resource conservation, the preservation of agricultural land, ensuring the availability of housing and employment for all Vermont's citizens, maintaining the social diversity and vitality of communities and ensuring the economic strength of communities. The statement was also required to 'reflect the shared values of Vermonters as expressed to the Commission in the course of its work' (Kunin, 1987,p.1). The third task required of the Commission was that it should make recommendations on how to use the goals and principles to guide state, regional and local decisions and that the recommendations of the Commission 'should discuss the applications of the goals and principles to financial assistance, public investment, development regulations, and the public planning process' (Kunin, 1987,p.1).

After a fourteen-month process which included eight public hearings around the state, the Commission developed a comprehensive set of guidelines for growth and recommendations for a process to guide the future pattern of development in Vermont. These guidelines for growth encompass a large number of specific recommendations on planning, economic development, natural resources, agriculture and affordable housing which are intended to maintain the traditional village-town-countryside settlement pattern which reflects the essence of Vermont. The Vermont ethic is that there should be a clear delineation between town and countryside and that this principle should form the basis of all plans developed by state, regional and local planning authorities. The four broad strategic goals identified to guide future policies and planning at all levels of Vermont government were: to maintain a sense of community; to maintain Vermont's agricultural heritage and working landscape; to protect environmental quality; and to provide the opportunity for all state residents to obtain a quality job, good education and decent, affordable housing.

In designing the process for shaping Vermont's future, the Commission took into account several points which were taken directly from the public hearings. State residents wanted decisions made at the lowest possible level of government and did not want additional layers of government. Residents recognised that the changes which occurred in one locality would often affect more than one community and that most municipalities did not have adequate financial, technical and legal resources to manage growth effectively. It was also recognised that Vermont did not have a clearly defined process for dealing with developments which would affect several towns, that effective decision-making through local and regional planning would help to guide public investment decisions and reduce the burden on the regulatory process and that state agencies often had conflicting mission statements which interfered with their ability to coordinate and integrate their planning and development activities effectively.

The resulting process recommended by the Commission was predicated on the belief that comprehensive plans must be developed at the local, regional and state levels and that the plans should be integrated with each other. The process as stated by the Commission included

several discrete stages: it was recommended first, that the planning
process would begin with the adoption of guidelines which would provide
a policy framework for developing more detailed plans; second, that
land-use data would be acquired and that regional information needs
would be identified; third, that town, regional and state agency plans
should then be developed concurrently and be based on this common
information; and, fourth, that Regional Planning Commissions would
then review the town plans to ensure consistency with the basic guidelines
and with regional needs. It was also recommended that a Council of
Regional Commissions should be formed to review regional and state
agency plans (Costle *et al.*, 1988, pp.16-19). It was from the second and
third recommendations (to gather land-use data to identify regional
information needs and to integrate planning activity) that the Vermont
GIS drew much of its political mandate and support.

The Growth Management Act of 1988 (Act 200) emerged as a direct
result of the Commission's work and was designed primarily to
strengthen the process of integrating planning between the local, regional
and state levels. As part of this comprehensive planning package,
US$4.75 million was appropriated for the development of a state-wide
GIS over a five-year period. The annual appropriation of approximately
US$1.0 million is to be generated by a new real estate transfer tax
implemented by the state. Although the annual appropriation to the
Vermont GIS is currently guaranteed for only five years, it is expected
that upon completion of the five-year development phase, a permanent
appropriation will be tied to the real estate transfer tax.

The University of Vermont: an advocate for GIS

By the time the Governor's Commission on Vermont's future had
finished its report in 1988, the School of Natural Resources at the
University of Vermont had already established a long and successful
track record with GIS. For almost ten years the School of Natural
Resources had been developing a GIS Laboratory of national
prominence. For example, the laboratory at the University of Vermont
was one of the first ARC/INFO platforms in the United States, and in fact
was the first successfully to port ARC/INFO to the DEC family of VAX
minicomputers.

In addition to being one of the early platforms for topologically-based
vector GIS software, the laboratory had already applied significant
resources to demonstrating the value of GIS for planning. The staff at the
laboratory had had an involvement with local and regional planning
commissions in demonstrating and educating planners to the value of

GIS. Long-term GIS data base development efforts had been commenced by the laboratory for several Vermont towns as early as 1980: all of these continue today. The contribution of the University of Vermont should be seen as a critical element in the development of the Vermont GIS for the following reasons: first, the staff of the School of Natural Resources had recognised very early the value of GIS and had invested significant time, energy and money in developing technical expertise in the hardware, software and data base issues which underlie it; second, the staff of the GIS Laboratory were an active force in transferring the technology of GIS from an academic laboratory environment to public-sector planning, and, third, the activities of the University's GIS Laboratory over an almost ten-year period had resulted in the development of significant cartographic and attribute data bases for many communities throughout Vermont.

The foresight and efforts of the University of Vermont can be seen as an important contribution to the development of the Vermont GIS since by the time the political motivation and commitment for a state-wide GIS were in place, the University had already developed much of the technical expertise, the resources and the application experience needed: it had also developed several major planning-relevant GIS data bases. The activities of the School of Natural Resources at the University of Vermont continue to be one of the important foundations of the Vermont GIS.

Design and organisation of the Vermont GIS

Because the Vermont GIS is intended to support planning activities at the state, regional and local levels of government, it was necessary to develop a delivery model for products and services which facilitated coordination between the different planning levels, while at the same time ensuring adequate access to all of the state's 250 cities and towns. The fundamental problem in developing a state-wide delivery model for GIS products and services was that coordination between all users was necessary to develop, maintain and safeguard the quality of GIS data bases. The need for this type of coordination favoured a central service bureau model, where all processing would be done at a single site. However, such a central service bureau can restrict efficient access to GIS products and services. For example, towns which are located at a greater distance from the service bureau may be at a disadvantage compared to towns which are closer, since the formative stages of developing most GIS applications require a great deal of contact between planners and GIS experts.

Additionally, a central service bureau can have a greater problem in remaining responsive to the needs of the user community, since it is often required to operate on a 'first come, first served' basis, which, although seemingly fair, seldom results in an effective or equitable distribution of services.

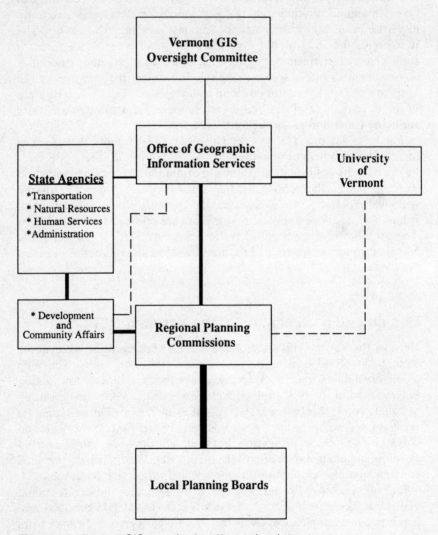

Figure 4.1 *Vermont GIS organisational/operational structure*

The alternative solution is to have a network of processing sites which provides better access to towns, but which increases the problems associated with training operators, maintaining multiple hardware and software installations and, most important, ensuring the quality and integrity of the GIS cartographic and attribute data bases. The delivery model developed by the Vermont GIS is both ambitious and unique: it is ambitious because it attempts to provide multiple and geographically distributed delivery platforms to provide the best access possible to local planning agencies and it is unique because it has embedded within it a management structure which facilitates coordination between the multiple delivery platforms and a central state data clearing house. An overview of the organisational structure of the Vermont GIS is shown in Figure 4.1.

The delivery of GIS products and services for local planning

The delivery model for GIS products and services for Vermont cities and towns has been designed to use the state's twelve regional planning commissions (RPC) as platforms for GIS. Each of the twelve RPCs would be equipped with computer hardware, software, data bases and GIS staff in order to provide a complete array of GIS services for each of the RPC's constituent communities. There were two prime reasons for adopting this type of delivery model for local planning enterprises. The first reason was derived from the fact that it had been determined that the planning philosophy in Vermont should be a bottom-up rather than a top-down process and so it was necessary to have a regional network of delivery platforms rather than a single central service bureau. The second reason was that the twelve RPCs provide a viable network of regional platforms since they offer relatively easy access to most Vermont towns. For example, the RPCs already had strong relationships with most towns in their regions since they are the primary source of technical planning expertise for local planning boards. RPCs also maintain strong relationships with state activities via the Agency of Development and Community Affairs, which is an important source both of technical planning expertise for the RPCs and of funding for both regional and local planning activities.

In addition to providing a regionalised distribution network of GIS platforms, which takes advantage of a pre-existing network already familiar to the planning community, the RPC delivery model was thought to provide the best opportunity for citizen participation in the GIS process. It was hoped that, by placing GIS platforms in RPCs, the GIS staffs could do outreach and education workshops for local communities and, as a result, citizens could become involved in the massive data capture process required to develop local attribute data bases. After a year of operation, the Vermont experience has been one of remarkable success both in having RPCs deliver products, and in involving local citizens in the data capture process through educational outreach (Millette, 1990).

The delivery of GIS products and services for state planning

The delivery model for GIS products and services for state planning is to equip each of Vermont's five major state agencies (Agency of Natural Resources, Agency of Transportation, Agency of Development and Community Affairs, Agency of Human Services and Agency of Administration) with a complete GIS delivery platform similar to those in the RPCs. For example, both the Agency of Natural Resources and the Agency of Transportation are actively involved in developing their GIS operations. As time and resources permit, the remaining agencies will develop their GIS operations over a four-year period.

The GIS operations in the agencies are intended to serve two basic functions. The first is to support agency-wide planning operations over the entire state. Since the agencies have responsibilities within their mandate which cover the entire state, GIS provides an ideal tool for identifying and analysing the geographical components of planning concerns. The second function of these state-level GIS platforms is to provide a common medium for the exchange of data between the state level planning activities and the local planning activities. Once state, regional and local GIS data bases are fully developed, the opportunities for sharing data between the various platforms is unlimited, and in fact are explicitly called for in the Governor's Growth Commission Report (Costle *et al.*, 1988, pp.18).

A continuing role for the University of Vermont

The School of Natural Resources at the University of Vermont has continued to play an important role in the Vermont GIS. Of paramount importance is the expertise that the staff has acquired with ARC/INFO which is the software standard adopted by the Vermont GIS. Having access to this extensive knowledge base has been essential to the installation of ARC/INFO in the RPCs and state agencies. The University conducts regular training programmes for planners in the use of ARC/INFO. The University also plays a major role in data capture for the state-wide data layers. For example, the laboratory is currently under contract to the Office of Geographic Information Services to digitise the

US Soil Conservation Service's soil associations for the entire state. The contribution of the University's School of Natural Resources with regard to the Vermont GIS has so far been as prime instigator and supporter. Were it not for their technical competence, active participation and role in facilitating the process of GIS technology transfer, it is unlikely the Vermont GIS would have ever got off the ground.

The Office of Geographic Information Services

The Office of Geographic Information Services (OGIS) is a newly created entity within an existing state agency which plays a pivotal role in keeping the Vermont GIS truly state-wide in its activity. If one considers the problems inherent in a state-wide GIS which will ultimately have twenty individual processing platforms, about sixty professional and technical staff and numerous cartographic and attribute data bases in varying stages of development, it can be seen that this will pose daunting problems particularly in terms of overall coordination. Specifically, there are problems with respect to data which include developing a comprehensive set of data standards, ensuring that data standards are respected and the acquisition and distribution of large GIS data bases. The task of developing a set of data standards which will adequately suit the needs of planners at the state, regional and local levels is no simple task. The problem of ensuring that data capture and processing activities at twenty individual sites scattered around the state faithfully adhere to the state data standards is even more difficult. The problem with data standards is that they slow down the data capture process and discourage the use of 'quick and dirty' analyses. However, without data standards, important characteristics, such as spatial resolution and the genealogy of data layers, can be obscured, making the value of analyses done with these data layers of dubious value.

The Vermont OGIS plays a central role in all aspects of data standards, data distribution and many other important coordinating functions. For example, with the assistance of staff from the School of Natural Resources at the University of Vermont, OGIS developed a comprehensive set of data standards for the Vermont GIS and actively works to ensure that all state and regional GIS platforms adhere to these standards. As the administrative head of the state GIS, OGIS enters into contracts with a variety of public and private concerns for GIS-related hardware, software, training and data capture. The OGIS runs a monthly GIS users group meeting which allows staff from the various GIS platforms to network and share learning experiences. The OGIS runs an annual Vermont GIS Conference to educate town planning boards and officials about GIS and to demonstrate its uses for local planning. The OGIS plays an active role in lobbying both the governor's office and the state legislature for continued support in addition to the annual appropriation provided by the real estate transfer tax.

The role of the OGIS is both pivotal and essential because it is at the same time part manager and part partner in the state GIS process. It is so positioned that it sits in the centre of all GIS activities. Because it has daily contact with the RPC platforms, the state agency platforms, the laboratory at the University of Vermont and the political activities in the state capital, OGIS has become an effective broker and clearing-house for GIS data, knowledge, experience and funding. Most important, OGIS provides the coordination between the various activities so essential to a state-wide GIS with such ambitious goals.

The Vermont GIS Oversight Committee

Although the OGIS functions as the day-to-day administrative head of the Vermont GIS, there is a fourteen-member Oversight Committee which meets monthly to set GIS policy and to monitor the progress of the various GIS activities. The committee is made up of representatives from local planning boards, regional planning commissions, state agencies, the state university and the private sector. The intention behind the creation of the Oversight Committee was to ensure that the Vermont GIS remained responsive to the needs of all planning agencies throughout the state and to prevent it from becoming a self-serving government bureaucracy. To this extent the Oversight Committee has been extremely successful. To date the Oversight Committee has developed a number of important policies including the decision to select a single GIS software package (ARC/INFO) as a state supported standard to optimise the efficiency of data capture, data sharing, user training and user support; the decision to develop individual GIS data bases for local and state-wide analysis (the local planning data base is based on Vermont's 1:5000 orthophotos while the state-wide planning data base is based on the USGS 1:100,000 Digital Line Graph (DLG) files); and the decision to keep the Vermont GIS under local democratic control. Since its creation, the GIS Oversight Committee has received several offers from commercial enterprises to subsidise the enormous data capture effort that the state is presently undertaking and has estimated to cost in excess of US$37,000,000 over five years (Crosswell and Archer, 1989,pp.5.17-5.20 and Kunin, 1989,p.10). This would be in exchange for rights to manage the Vermont GIS and to market its products and services. Although the idea of having the data capture effort subsidised is appealing, in the long run the Oversight Committee felt it would probably cost the state more to pay commercial rates for data and services than they would be likely to save by such a subsidy.

The GIS Oversight Committee has effectively complemented the OGIS, in that while OGIS has focused its efforts on dealing with the day-to-day activities of the GIS platforms, the Oversight Committee has focused on the strategic issues and on ensuring that the philosophies which led to the development of the programme have not become obscured by more mundane hardware, software and data base issues.

The Vermont GIS planning data bases

Since one of the guiding principles behind the growth management process in Vermont is to allow decisions to be made at the lowest possible level of government, it has become necessary to develop GIS data bases which will support not only state-level planning, but also local-level planning. The application priorities for the Vermont GIS can be seen in Table 4.1 and it is clear that these application priorities explicitly include both state-level and local-level applications. As a result, the decision was made to develop two distinct sets of data bases — one for state-level and one for local-level planning. The state and local data bases each contain individual cartographic and attribute elements which were identified as being most useful for the typical types of applications likely to be required by planners. These cartographic and attribute elements were identified from a survey which was part of a comprehensive GIS strategy study conducted by a private consulting firm and are reported in Croswell and Archer (1989).

The state-level data base

The state-level data base is spatially tied to the USGS Digital Line Graphs (DLG) at a scale of 1:100,000. These digital base maps were originally complied from 30×60 minute USGS quadrangles and include transportation route centrelines, hydrography and some boundary information. The 1:100,000 DLGs meet National Map Accuracy Standards and are accurate to within plus or minus 167.0 feet. The state-level data base is intended primarily to support the agencies which are required to undertake state-wide analysis and planning. For example, the Department of Health is presently mapping well-head protection areas for the entire state and putting them into the state GIS. The Agency of Transportation is updating the road and highway data base for the

Table 4.1 Vermont GIS application priorities

	PRIORITY 1	PRIORITY 2
Natural Resources	Flood plain mapping Wetland mapping Land cover mapping Sensitive natural areas Timberstand management Water discharge tracking Hazardous/Solid waste planning	Water well drilling inventory Environmental impact analysis Ground water modelling Air discharge mapping
Land-use planning/ zoning	Demographic analysis Historic site mapping Land -use map production Zoning map production	Comprehensive planning Permit/Development tracking Rezoning evaluations Redistricting analysis Public land acquisition
Public Works/ Transportation	Road system map production	Traffic volume/Accident study Route analysis Site plan reviews Right-of-way management Pavement management
Parcel Mapping/ Appraisal	Parcels in current use Tax mapping analysis Ownership searches/Query	Property map updating Special property mapping Field appraisal Scheduling/Routing appraisals
Water/Sewer utilities	Well/Septic location mapping	Water supply permit review Water/Sewer map updating Mining Sand/Gravel pits and rock Quarry mapping

Source: Kunin, 1989,p.2

entire state along with paving and maintenance histories and putting them into the state GIS. The current schedule for data capture into the state-level data base can be seen in Table 4.2.

Local-level data base

The local-level data base is spatially tied to a state-wide coverage of

1:5,000 scale orthophotos (1:1,250 in the urban areas) which is part of an
ongoing state programme for mapping land-use for tax purposes termed
the 'Current Use Program'. Since the Current Use Program has the entire
state captured in large-scale, ortho-rectified aerial photos with state
plane coordinates, it provided a logical data standard for the local

Table 4.2 State-level data base data layers

Priority 1: (by July 1991)	Priority 2: (by July 1994)	Priority 3: (after July 1994)
DLG Base maps Transportation centrelines Surface water Political/Administrative boundaries Census areas Land-use/land cover Wetlands	Soil associations Groundwater features Historical/ Archaeological sites Sensitive natural areas Topographic contours Watersheds.	Geology Recreational facilities Infrastructure facilities Solid/Hazardous waste sites Utility transmission lines

Source: Kunin, 1989, p.5

planning data base. The need for high levels of positional accuracy for
local planning applications is adequately served by the orthophotos which
are accurate to plus or minus 10.0 feet.

The local-level data base is intended to meet the needs of local planning
boards and regional planning agencies for local planning applications.
Typical applications of GIS to local planning include: town land-use/land-
cover inventories, parking studies, traffic flow studies, local LESA (land
evaluation and site assessment), studies for agriculture and forestry and
zoning evaluations. The current schedule for data capture into the local-
level data base can be seen in Table 4.3.

Vermont GIS data base concerns

The development of GIS spatial and attribute data bases for an entire
state the size of Vermont poses significant challenges. The most basic
questions about which types of data to include and to what levels of detail
they should be captured are not easy to resolve. Even more challenging is
finding the enormous amount of money required to finance such a large-
scale and quality-sensitive data capture effort. Once the data bases are
developed, where does the responsibility lie for managing, maintaining
and updating them? Who is allowed access to the data and how should the

data be distributed? Who should determine appropriate fee schedules for data and data processing? These are all problems that were encountered

Table 4.3 Local-level data base data layers

Priority 1: (by July 1991)	Priority 2: (by July 1994)	Priority 3: (after July 1994)
Orthophoto Base Map Property Parcels Transportation Centrelines Land Use/Land Cover Zoning Flood Zones Political/Administrative Boundaries	Soil Associations Sanitary Sewer Facilities Water Facilities Septic Systems	Topographic Contours (2.0 foot interval) Storm Water Facilities Dams, Bridges, Culverts, Etc.

Source: Kunin, 1989, p.5)

almost immediately upon the first GIS pilot platform being put into operation in March 1989. It would be less than honest to give the impression that the Vermont GIS has totally resolved all these concerns. However, due to the unique philosophy and administrative structuring of the Vermont GIS, it has been possible to develop both short-term and long-term plans for dealing with many of these very practical issues.

For example, a GIS data base plan has been developed which takes both a coordinated and flexible approach to the long-term data capture process. A comprehensive set of data standards has been adopted by the GIS Oversight Committee and the responsibility for their enforcement has been placed with the OGIS. As a result, data captured by any of the state agencies or RPCs can be incorporated directly into the state GIS data bases providing the data meet the state data standards. The OGIS has developed a number of programmes to finance the acquisition of key data layers. The first programme is a contract with the University of Vermont to digitise soil maps for the entire state published by the Soil Conservation Service. The second programme pays for those towns which have parcel maps to digitise them into ARC/INFO format. For towns which have not yet been parcel-mapped, OGIS has tried to assist them with advice on the benefits involved and will provide a modest subsidy towards digitising them into ARC/INFO should a town decide to undergo the process. Probably the clearest evidence of the flexibility of the data base development plan is that OGIS will consider participating in data capture projects on a cost-sharing basis with any Vermont community as long as the project is considered to be involved with growth management, the community agrees to follow state GIS data standards and OGIS has funds available in its data capture programme.

With regard to access and the pricing of data, the Oversight Committee has determined that any public concern, private concern or citizen can have access to any data in the state GIS at the cost of duplication. However, this cost of duplication includes the media cost and the real cost of a GIS technician's time to download the raw data. The policy for RPCs pricing GIS processing and downloading services is currently such that RPCs can charge what they feel is a 'fair market' price based on their need to recover the cost of GIS-related salaries, hardware, software and training. All of the presently operating RPC GIS platforms charge between US$25.00 and US$35.00 an hour for GIS services which is considerably less expensive than most current commercial rates for similar services.

GIS in the Regional Planning Commissions

One of the most unusual characteristics of the Vermont GIS is the way it uses RPCs to provide GIS products and services to local planning enterprises. The specific responsibilities of the RPCs with regard to the state GIS fall into three general categories which include providing technical advice and education for local planning enterprises, serving as regional production and distribution facilities for data and maps and providing the OGIS with information and recommendations concerning the types of data and services most useful for local planning activities. The role of providing technical advice and education to local planning agencies is essential since few of them have any direct experience with GIS. The education of local planners to GIS should help create the market for products and services. As regional platforms for GIS, each RPC will maintain a fully equipped laboratory facility with microcomputers, ARC/INFO software, local cartographic and attribute data bases, a large format digitiser, an E-size plotter and trained staff. Although each RPC will have a large format digitiser, they are not intended to become large-scale data capture operations. It is intended that RPCs will concentrate on integrating GIS into planning activities, processing data for analysis and delivering graphic products: large-scale data capture projects will be contracted to commercial vendors. The staffing of GIS activities in the RPCs will be done by a planner who will double as GIS manager and by at least one full-time GIS technician. As counsel to the OGIS, RPCs play an important role in ensuring that the state GIS remains responsive to local planning needs and activities.

GIS application in the Regional Planning Commissions

In March 1989 the Vermont GIS began a series of pilot projects which were intended to test the ability of RPCs to deliver GIS products and services to local towns. Having been in operation for almost a year, these pilot projects would be better described as an assortment of GIS activities in support of planning, rather than as discrete GIS projects. These activities include data layer development for local planning purposes, data layer development for regional planning purposes and data layer analysis for specific problems thought to be indicative of local and regional planning needs. Only one of the five RPC GIS platforms currently running is applying GIS exclusively to a single specific problem — that being the use of GIS to support the location of solid waste disposal facilities.

Local planning applications

The integration of GIS into local planning activities has included a wide range of activities which are too numerous to report here in detail. However, three typical examples include a community-wide analysis of prime agricultural soils, an updating of local flood plain location data and the development of a comprehensive town parcel layer.

Several of the pilot RPCs have had remarkable success with integrating GIS into planning. For example, when the town of Woodstock wished to develop criteria for the protection of prime agricultural soils, it was necessary to produce an inventory of these soils in order to determine which had remained available for agriculture and which had already been developed upon. The analysis was done by taking the town parcel layer and identifying the parcels and portions of parcels which had sustained development in some manner. This development layer was then overlayed on to the town's soil layer which had been digitised from Soil Conservation Service maps. The resulting analysis yielded a map and tabular acreage summaries of total prime agricultural soils, prime agricultural soils which have been lost to development and undeveloped prime agricultural soils which are available for protection. Since the Woodstock soil and parcel layers had already been captured into the state GIS, the total effort for this analysis was less than two person-days.

A second example of the type of activity likely to be encountered in local planning applications is the continual process of updating and improving GIS data bases. In the course of developing a flood plain data layer to be used for zoning in the town of Woodstock, Flood Insurance Rate Maps (FIRMs) produced by the Federal Emergency Management Agency (FEMA) were compared to the town's existing hydrology data layer. The comparison showed that the fit of the Woodstock hydrology layer and the FIRMs were very close except that a number of features present on the FIRMs were missing from the hydrology layer. Further research determined that the features present in the FIRMs were in fact accurate and so a decision was made to add them to the hydrology layer. Subsequent analysis determined that the reason for the absence of the features from the Woodstock hydrology layer was that it had been compiled from 1:5,000 scale orthophotos, while the FIRMS were compiled from 1:2,400 scale survey maps. In this case, the process of developing a new data has led to the improvement of an existing data layer at a relatively low cost.

A third local planning application conducted by an RPC in Central Vermont involved developing a comprehensive parcel data layer for the town of Calais. This parcel layer included a digital map of the town's approximately 1,100 parcels and an extensive parcel attribute data base. Of particular interest with regard to the attribute data base was that it was developed by merging the town assessor's grand list (local tax roll) with the state's CAPTAP (Computer Assisted Property Tax Assessment Program) data base and relating them to the parcel map by parcel identification numbers. The resulting data base now includes a large number of attributes for each parcel including owner, acreage, current use, number and types of dwellings, non-dwelling structures and many other data items relevant to planning and growth management.

What makes the development of the Calais parcel layer of special significance is that it represents an unusual cooperative effort between the state, the regional commission and the municipality to develop a product with a low level of data conflict which can be utilised by both local planning and municipal administrative concerns. It should also be mentioned that the CAPTAP and grand list data bases were consolidated and linked to the parcel map as an ARC/INFO coverage using the PC version of the software. Although the consolidation of the attribute data bases into INFO files was not easy or straightforward (it could not be done as a simple export/import routine) the entire development including digitising the parcels took less than 150 person-hours.

Regional planning applications

Regional planning applications to date have typically lagged behind those of local planning primarily due to the shortage of regional data bases. The shortage of regional data bases is the result of the limited time that the state GIS has been operational and the fact that the bulk of the funding for data capture has been for local data base development in scattered rather than contiguous areas — this has tended to preclude their use for regional applications. However, one RPC is presently using GIS to help site a regional solid waste disposal facility.

The criteria presently being used by the RPC for suitable landfill sites includes soil permeability (based on the Soil Conservation Service soil classification), parcel size to ensure the facility can be of sufficient size to justify its development, depth to bedrock and depth of water table to ensure that decomposition will occur without the contamination of groundwater. Although there are several sources of guidelines for these criteria (e.g. from the State of Vermont, the Environmental Protection Agency and the Soil Conservation Service), there is no consensus between the various guidelines on the best conditions for locating landfills. Potential sites were identified by the RPC by selecting soil polygons that had suitable permeability characteristics and which were of sufficient size to justify further investigation.

Prior to the GIS being put in place, a team of local planners identified six potential sites for landfills from visual inspection of Soil Conservation Service Maps which required a total effort of approximately 18 person-days. Once the GIS was operational and the maps digitised, the same maps yielded 14 sites in several person-hours. The use of GIS to date has been restricted to selecting candidate sites for landfills based on a soil polygon's permeability characteristics and size. However, once the ultimate selection criteria are better defined, the GIS can be used to examine additional factors such as ownership, zoning, land use, agricultural potential, transportation access and appraisal value. In this manner, a more complete, more rigorous and more cost-effective analysis can be made of the candidate sites than if traditional, non-GIS techniques were used.

A second regional application currently under review is the use of GIS in conjunction with the processing of satellite produced images to develop up-to-date land-use data layers for inclusion into regional and (in limited cases) local plans. Although the Vermont 1:5,000 series orthophotos provides a complete collection of very high resolution photography which is ideal for mapping land-use, many of the photos are now ten to fifteen years old. Given the rapid growth and development that has taken place in Vermont during this period, their value for mapping land-use has been compromised. The use of image processing of SPOT satellite imagery, combining it with other GIS layers for updating regional land-use data layers, is currently being investigated. It is not anticipated that RPCs will develop image processing capabilities. However, it is possible that the OGIS or the University of Vermont could develop a production capacity for classification of SPOT imagery for land-use data layers which could then be distributed to RPCs.

Although the pilot study is not yet complete, early indications are that the process, given rigorous quality control, is accurate and has distinct advantages over commissioning more frequent aerial photography or conducting traditional windshield surveys to map land-use in rapidly developing areas. The key to developing accurate land-use data layers from image processing of SPOT imagery lies in the development of the land-use/land-cover classification system. Planners wishing to make use of SPOT imagery for regional land-use must follow two simple but important rules. The first rule is that when choosing the specific land-use/ land-cover classes to be included in a regional GIS data layer, planners should make sure that the classes selected will ultimately provide information which will be useful in planning. It is a mistake for a planner to model a classification system after something seen in the literature if it has no particular relevance to his particular planning agenda. The second rule is that it is important to ensure that the classification system developed for the imagery analysis is compatible with the regional land-use map to be included with the regional plan. All too frequently planners expect satellite imagery to yield significantly more detailed classifications than they ever intend including in the regional plan and this commonly leads to increases in the errors of commission. The advantage of using satellite imagery and image processing rather than custom-flown aerial photography and visual interpretation is that it is less expensive to acquire, less expensive to classify and, because it is already in digital form, much less expensive to incorporate into a GIS. The advantage of satellite imagery analysis to windshield survey is that land-use classifications for large areas are more complete, more consistent and can generally be done for less money and in shorter time periods.

Vermont GIS application areas: planning-led not technology-led

Given the nature of the GIS applications discussed above, it is clear that the Vermont RPC GIS platforms are not working on the cutting edge of GIS application innovations. The types of activities which consume most of the time and resources of the GIS staff in the RPCs are fairly basic. Developing local parcel-based data bases and processing them to develop basic but essential products for use in zoning, land-use, transportation and natural resource analyses, presently keeps the RPC GIS staff operating at full capacity. However, the recognition of this fact is itself extremely important. In the recent explosion of activity and interest in GIS, it can sometimes be difficult to separate fact from fiction. It is possible for the glittering array of possibilities to obscure the pragmatic sensibilities required to put the technology to effective use. From the perspective of local and regional planners, GIS is a tool which integrates and hopefully simplifies the processes of mapping, data base management and spatial data analysis. It is a tool to support the process of planning and therefore the applications should reflect the planning agenda at hand. If GIS activities in an RPC are allowed to become activities in support of the technology rather than in support of the planning agenda, then the GIS implementation becomes counterproductive to the RPC's mandate. Obviously, the implementation of a GIS in an RPC is going to include financial, administrative and personnel overheads but it is important that these overheads are kept in balance so that in the final analysis, resources are going to support planning rather than simply to support the GIS.

The Vermont experience so far has been one of remarkable success. With only a few exceptions, the vast majority of RPC applications of GIS have been very successful. The success of these applications can be credited to three important factors: first, that the RPC GIS platforms have been equipped with adequate hardware, software and data to enable them to produce useful planning products; second, that the staff of the RPC GIS have been given adequate institutional support from both the RPC management and the OGIS so that they have adequate time, training and motivation to succeed with GIS; and third, that the staff of the RPC GIS are firmly grounded in planning and understand the role that GIS can play in support of the planning agenda.

If there is a GIS cutting edge that the Vermont GIS can claim, it may be that it has developed a model for a state-wide GIS which is extremely ambitious and that seems to work. Of critical importance to the success of the model is the coordination required between the key players in the GIS delivery process. It is precisely because there is coordination between RPCs, the OGIS, the University of Vermont and the GIS Oversight Committee that the Vermont GIS works as well as it does. Although it is still early in the implementation stages of the system, preliminary indications are that the Vermont GIS is making a valuable contribution to planning for growth management in Vermont.

Conclusions

The Vermont GIS represents one of the most unusual approaches to a state-wide GIS delivery model yet seen in the United States. Its development and implementation is the product of a great deal of cooperation between local, regional and state levels of government. The success to date of the Vermont GIS can be attributed to three main factors. First, the Vermont GIS was derived from a political process and designed to meet clear planning needs: as a result of this, from its inception it had a clear political mandate and a secure source of funding. Second, the Vermont GIS has been developed in response to the growth management planning needs at local, regional and state levels of government and this recognition of planning needs at all levels of Vermont government has led to a philosophy of cooperation among GIS advocates throughout the state. Third, the developers of the Vermont GIS recognised early in the development phase that it would take more than computer hardware and software to make the GIS successful: it would require expertise in technical planning, education, management, coordination, marketing and lobbying in addition to the technical issues of GIS to create a viable and successful state-wide GIS in the State of Vermont.

Although there are a growing number of state GIS programmes of varying types currently under development in the United States, none has yet taken as ambitious an approach as the Vermont GIS. The success experienced so far by the Vermont GIS can be credited to the integrated and coordinated approach taken to the technical, administrative, financial and political issues facing the development and implementation of the state-wide GIS system.

Bibliography

Costle, D.M., Wheeler, B., Bradley, D., Gibb, A., Tarinelli, D., Patenaude, W., Bankowski, E., Snelling, M., Ryan, J., Jensen, M., Lawson, R. and Billings, P. (1988), *Report of the governor's commission on Vermont's future: guidelines for growth*, Vermont Office of Policy Research and Coordination and the Vermont Agency of Development and Community Affairs, Montpelier, Vermont.

Croswell, P.L. and Archer, A. (1988), *State of Vermont GIS needs determination: final report for phase I of Vermont's comprehensive GIS strategy study*, PlanGraphics, Inc., Frankfort, Kentucky.

Croswell, P.L. and Archer, A. (1989), *State of Vermont conceptual system design: final report for phase II of Vermont's comprehensive GIS strategy study*, PlanGraphics, Inc., Frankfort, Kentucky.

Kunin, M.M. (1987), *State of Vermont executive order*, 22 September 1987, Montpelier, Vermont.

Kunin, M.M. (1989), *Vermont Geographic Information System: annual report to the legislature,* Office of the Governor, Montpelier, Vermont.

Millette, T.L. (1989), 'Implementation of a state-wide GIS: Regional Planning Commissions as platforms for products and services', *Proceedings of GIS/LIS '89,* Orlando, 2, pp.438-47.

Millette, T.L. (1990), 'GIS in the Regional Planning Commission — expectations and exasperations: case studies from the Vermont State GIS', *GIS for the 90s: Proceedings of the Second Canadian National Conference on Geographic Information Systems,* Ottawa, Canada (forthcoming).

Chapter Five
The development and application of Geographic Information Systems in Hawaii

Karl E. Kim

Introduction

To many, Hawaii is perceived as an island paradise somewhere in the middle of the Pacific Ocean. Yet to those interested in geographic information systems (GIS), Hawaii may be much more than a vacation spot or an ideal setting for an international conference. Hawaii has a long history of interest and involvement in GIS technology. In the mid-70s, for example, the University of Hawaii developed a prototype GIS known as the Geographic Information and Display System (GIDS). In the late 1970s, the State of Hawaii engaged in a number of cooperative demonstration projects with the National Aeronautics and Space Administration (NASA) using LANDSAT data to map land use, land-cover and water-cover classifications in rural and urban areas. In recent years, both state and local governments have initiated programs greatly to enhance both computerised mapping and data management capabilities. Efforts to digitise most of the widely used maps in the state are well under way. So, too, is the long and tedious process of building attribute files on everything from population, buildings, zoning, endangered plants and animals, coastlines, historic and archaeological sites and other important environmental aspects of Hawaii.

For a small state, Hawaii has big plans for GIS development and for the application of GIS techniques to practical planning and decision-making situations. This chapter is concerned with the development of GIS in Hawaii and points to some of the major issues, problems and opportunities which have recently arisen. Efforts undertaken by the State of Hawaii and the City and County of Honolulu are described and the differences between their respective approaches are noted. GIS in Hawaii is viewed as a future investment with certain risks and potential pay-offs. At the same time, efforts to build an operational GIS system may have some additional benefits which go beyond those initially identified. Data sharing, resource pooling and the joint management of

GIS across public agencies may produce new efficiencies and improvements in both routine and non-routine government operations.

In many respects, Hawaii is an ideal setting in which to explore the diffusion of ideas, the practicality of inventions and the success or failure of planned interventions. There are approximately one million residents in the state, with more than 830,000 people living on the island of Oahu. Land in Hawaii is scarce and under considerable pressure for development. Half of the 10,380 square kilometres in the state is zoned as conservation land, not only because of steep volcanic slopes, but also because of the need to preserve watersheds. In addition to its geographic isolation, Hawaii has a relatively simple and highly centralised system of government: there are only four county governments (Honolulu, Maui, Kauai and Hawaii) and one state government.

Perhaps more important and relevant to the application of GIS in a practical planning situation is the fact that planning, environmental management and resource allocation decisions are extremely critical in an island setting such as Hawaii. The population since statehood in 1959 has more than doubled. The tourism industry since the mid-1960s has mushroomed to the point where more than 5.7 million tourists now come to Hawaii each year. The growth in population has strained what is a very sensitive natural ecosystem and has increased demands on the public infrastructure. Agricultural activities (the cultivation of sugar-cane and pineapple) have been replaced by service industries such as tourism, leisure and commercial activities. There have been corresponding changes in land-use as more and more agricultural land has been converted to urban, tourism-orientated and commercial uses. At present, there are numerous large-scale development and redevelopment plans under way in Hawaii.

Because of the pace of development and the associated burdens imposed on government, the need for information has increased greatly in recent years: it is a trend which is also likely to continue. At the local level of government, the number of permit applications, zoning changes and requests for development approvals have increased substantially. For instance, the number of housing units has increased from 216,774 in 1970 to more than 385,290 by 1989. A reflection of the attractiveness of Hawaii is that the number of condominium units has grown from 15,320 to over 97,000 in the same period. Over the past two decades, Hawaii's gross state product in real dollars has more than doubled and the per capita personal income in the state is the fifteenth highest of the fifty states in the USA. This prosperity has helped to produce a state surplus in the past year of approximately half a billion dollars and unprecedented levels of revenue collection in the county governments despite cuts in federal funding. The growth and development in Hawaii have meant that not only has there been an increase in the need for information to support planning activities, but also there have been more government resources available to expend on the acquisition and development of new technologies such as GIS.

Hawaii's unique geography

Part of the reason why GIS technologies have so readily captured the attention of government officials in Hawaii is the state's attractiveness as a place to live, its unique geography, its unique and highly sensitive ecology and the resulting pressure to manage prudently land and other precious resources. The islands are almost wholly volcanic and are part of a chain of small islands and atolls which arc across a vast stretch of the Pacific Ocean. While there is still substantial volcanic activity on the island of Hawaii, most of the other islands remain relatively stable. Over the past thirty million years, the ocean and streams have greatly eroded the landscape, sculpting dramatic cliffs and valleys in the sides of volcanic mountains and creating the spectacular scenery that has become synonymous with Hawaii.

Hawaii's geography is such that there are relatively few places suitable for habitation and the most attractive of those areas, principally along the coastlines, are subject to a variety of natural destructive forces — hurricanes, flooding, *tsunami* (tidal waves) and coastal erosion. In addition, the climatology on each island is remarkably diverse. The windward and central mountainous areas of the islands have more precipitation than the low-lying, leeward and coastal parts of the state. The climatology and topography have helped to produce remarkably diverse plant and animal life. In addition to tropical rain forests, there are also lava fields with desert-like conditions, woodlands, pastures and forests which are home to some 2,500 kinds of higher plants which occur in Hawaii and nowhere else (Armstrong, 1983, p.69). In addition to many rare and endangered plant species, Hawaii's geographic isolation has made possible the evolution of many unique birds. Rare, indigenous species are constantly threatened by the introduction of new and destructive species of plant and animal life. Grazing animals as well as numerous pests continue to create problems for the state in its management of natural resource areas.

In 1961, Hawaii became the first state in the USA to adopt a state general plan and to enact a state-wide land-use law. This law established a land-use commission which classified all land in the state into one of four categories: urban, rural, agriculture or conservation. Urban districts include those lands already in urban use with a reserve for foreseeable growth. Rural districts are those lands principally used by small farms, although some low density residential development is allowed. Agricultural districts are those lands which are in agricultural production or have the potential for intensive cultivation. Conservation districts are those areas which have a general slope of 20 per cent or more and include forest reserves, water reserves, parks, marine waters and offshore islands. As of 1988, 166,507 acres in the state were classified as urban, 1.9 million acres as conservation, 1.9 million acres as agricultural, and 10,189 acres as rural (DBED, 1989, p.183). Under Hawaii's land-use control system, the county governments are responsible for administering land-use regulations in the urban areas through locally adopted zoning ordinances. For agricultural and rural districts, the state establishes regulations which the counties must administer (though counties may adopt more stringent controls if so desired). Conservation district lands are governed by the State Department of Land and Natural Resources.

The nature of Hawaii's land-use control system has several immediate implications for the application and development of GIS technology. First, it serves to divide state and county government responsibilities: counties are responsible for urban areas while the state is responsible for conservation, rural and agricultural areas. Second, because of the nature of each of these districts and the types of planning and resource management problems and issues which arise within them, the technical requirements for GIS vary by level of government. State concerns are more likely to focus on the broader resource management issues covering large geographic areas, while the counties are principally concerned with the management of development in urban areas, with zoning and the management and control of public facilities. Third, Hawaii's extensive system of land-use control suggests a proactive orientation towards planning and development within which management tools such as GIS may be inherently appealing. The fact that Hawaii is so isolated, that much of its economy is built upon tourism (which in turn depends upon the balance between the natural and built environment being sensitively managed) and that planners and planning are held in higher esteem (by virtue of strong land use laws and early attempts to regulate development) in the state than elsewhere, all helps to promote the use of state-of-the-art planning tools such as GIS.

The state perspective: resource management

Evidence of the importance of GIS to the State of Hawaii is contained in a resolution adopted by the House of Representatives in 1987 which asked the Department of Planning and Economic Development to chair a GIS Task Force and to recommend 'a GIS which provides maximum data with minimum redundancy while permitting various agencies and organisations, both public and private to access basic data, extract the information they need, and analyse it according to individual requirements' (H.R. no. 275, H.D. 1).

The Task Force issued a report in 1988 which recommended the development of a state-wide GIS system and the implementation of the minicomputer version of the ARC/INFO software package. The report stated that 'ARC/INFO which is already being used by the US Geological Survey, is emerging as a de facto standard for the US government and is also being used by many state and local governments' (OSP, 1988,p.17). With respect to hardware, the Task Force recommended the acquisition of a minicomputer central processing unit to be located at the Department of Budget and Finance's (DB&F) Electronic Data Processing Division (EDPD). EDPD was chosen because it was (and continues to be) the state's largest data processing centre and has the necessary facilities, resources and experience to maintain the state's GIS.

The Ocean and Coastal Information Management System (OCIMS) project

The approach taken by the state was to build a prototype GIS system which permitted the close inspection and evaluation of hardware and software before embarking on a more long-term process of identifying user needs, of completing the systems design, of developing data bases and applications and of training and educating users. The development of the state's GIS first began with the development and implementation of the Ocean and Coastal Information Management System (OCIMS). Following a brief survey of potential users of a state-operated ocean-related GIS in 1984, a pilot study for developing a GIS was conducted. In 1987, the state contracted with a private consultant to develop an ocean and coastal GIS data base. The firm was to evaluate the various hardware and software available, to select and purchase (for the state) the best system, to develop appropriate system programming, to design the OCIMS and to develop and demonstrate the data base.

The task of building a GIS to handle ocean-related concerns was a logical place to start, given the importance of the ocean resources and

Hawaii's long-standing involvement with coastal zone management (CZM). The OCIMS project was funded by the State Legislature and was developed largely to help carry out federal mandates including the development of management programs to achieve the wise use of the land and water resources in coastal zones, with consideration of 'ecological, cultural, historic, and esthetic values as well as needs for economic development'. The state CZM office serves in an advisory role, helping to coordinate actions of diverse agencies which have jurisdiction in coastal areas. These agencies include the Federal Army Corps of Engineers, the state departments of Business and Economic Development, Defence, Health, Hawaiian Home Lands, Land and Natural Resources and Transportation as well as several local agencies, including the various county planning agencies.

The state developed its GIS using the ARC/INFO system operating on a Prime minicomputer system. ARC/INFO was chosen because of its popularity throughout the USA and because both the US Geological Survey office in Honolulu and the City and County of Honolulu already used or planned to acquire this system. While there is other software available, ARC/INFO has become the market leader (see Maguire, 1989, p.174) and a virtual standard in GIS technologies in the USA. In justifying the selection of a minicomputer-based system, the Task Force determined that a microcomputer-based GIS would be inadequate for the state's needs.

The state's GIS operates on a Prime 4150, with 24 megabytes of main memory, augmented by two 770 megabyte direct access storage devices and a 1,600 bpi tape drive. Currently, 11 workstations (Tektronics 4111 and 4211) are connected either directly or by modem (9,600 bps) or through the state's local area network (19,200 bps). The majority of terminals are located in EDPD, but there are also terminals located in the Office of State Planning, the Office of Environmental Quality Control and the Department of Defence. Plans to install one terminal at the University of Hawaii are also under consideration. The GIS hardware also includes two Calcomp 9100 digitisers, a Calcomp 1044GT plotter and a laser printer. The total hardware and software costs amount to more than US$340,000. Maintenance costs are substantial, amounting to approximately US$24,000 for the hardware and an additional US$15,000 per year for the software. Some operating cost efficiencies have been realised by locating the GIS within EDPD which already had the personnel and the experience to manage such a minicomputer-based system.

The structure of the OCIMS data base is directly related to the way data are stored in the ARC/INFO system. Map information is stored in layers or coverages, which include the following features: arcs (lines), nodes (arc endpoints or line intersections), polygons (areas enclosed by arcs), tics (registration or geographic control points), label points and annotation. Attribute information is stored in three kinds of tables: polygons (areas), arcs (lines) and points. The combination of coverages and attribute data can be used to produce five different types of maps including the following: (1) area coverages (soils, parcels, population data etc.); (2) line coverages (utilities, infrastructure, streams, trails etc.); (3) point coverages (historic sites, wells etc.): (4) network coverages (combined polygon and line coverages such as bathymetry); and (5) linked layer coverages (for data with links between lines and points such as utility lines and fixtures). ARC/INFO utilises a relational data base management system which stores attribute data in flat, logical files where columns represent attributes and rows contain the attribute values for each feature.

The OCIMS data base is structured into seven map libraries. The first five libraries contain coverages and look-up tables organised around each of the five State Plan zones which also correspond to the major inhabited islands in the Hawaiian chain. Information at the parcel level is stored in these libraries and will include data such as the certified shoreline, location of Army Corps of Engineers permits and TMK (tax map key) data. The next library contains the US Geological Survey (USGS) base maps and will include primarily land-based information. In addition to containing data on elevations, flood zones, hydrology, rainfall, land cover, historic and archaeological sites and land-use, it will also include information on rare and endangered plant and animal species and traditional Hawaiian land divisions (ahupua'a). The final library will contain coverages on topics such as artificial reefs, bathymetric zones, fish catch areas, harbours and anchorages, navigational routes, marine ecosystems, recreation points, waste disposal sites and other ocean-related matters.

A number of potential GIS applications have been investigated and these include ocean waste disposal, accidental spills, beach erosion, maritime transportation, coastal energy facility development, ocean mining, ocean leasing, fisheries and marine conservation: however, it was found that much of the data needed for most of these applications were either unavailable or difficult to convert into a GIS format. Some of the applications, such as beach erosion, involve such complex, dynamic processes that they do not, according to some ocean engineers, readily lend themselves to being converted over to GIS: though this is an example of an area where developments in model-based simulation techniques are required if GIS is to be more effectively integrated into planning and management processes. Other topics, such as ocean mining, require the translation of vast amounts of technical data on water quality, sediment type and content and physical characteristics of nodules or crusts into a format suitable for GIS. This would require substantial participation on the part of experts and specialists. Much of the data that one might ordinarily assume is in a format suitable for GIS is, in fact, difficult to organise and standardise. For example, some data are collected with little or no locational references. Other information is aggregated in terms of very large geographic areas. While some data are organised temporally, other data sets are taken only at one point in time. Virtually all of the data that would be put into the OCIMS data base would have to be processed, reviewed, standardised, cleaned, checked and judged suitable for inclusion.

One proposed GIS application which was partially developed under the OCIMS project involved recreational water uses (fishing, harbour development and recreational activities). Each year an estimated ten million pounds in fish is caught in the state by recreational fisherman. The GIS could be used to identify the geographic distribution of fish populations to help maintain and manage the *'kapu'* system (a Hawaiian concept) of regulating (opening and closing) shoreline and nearshore fishing areas.

The State of Hawaii GIS

The OCIMS data base was used as a vehicle to demonstrate the potential of GIS. It has since served as one of the main foundations on which the state's system is being built. There are several strong justifications for this sort of process. It is important to note that the decisions reached were informed first by the GIS Task Force, a group of knowledgeable persons representing the major interested parties likely to take advantage of GIS. By limiting the scope of the GIS to ocean-related concerns, it gave decision-makers and others a hands-on opportunity to see something of what could be done with a GIS, to see how long it took to develop data bases and applications and to assess what the actual costs were. Finally, by actually purchasing the hardware and software, the state moved closer to the goal of building a state-wide GIS data base. Much of what was learned and acquired in the OCIMS project has been directly applied to the state's long-term GIS efforts.

The strategy adopted in the development of OCIMS was to rely heavily upon the USGS 7.5 minute quadrangle maps (7.5 minutes longitude by 7.5 minutes latitude, usually represented by a 1:24,000 scale) to serve as the base maps and then to rectify other maps, such as the Federal

Emergency Management Agency's Flood Insurance Rate Maps (FIRM)
or the state Department of Transportation's airport noise contour maps,
to be consistent with the USGS maps. By adopting the USGS quadrangle
maps as a base, all of the available data files produced by USGS and the
future Census Bureau TIGER (Topographically Integrated Geographic
Encoding and Referencing) files, as well as other data routinely
developed using the USGS base, could be easily integrated into the
system. Even before the implementation of GIS in Hawaii, the 1:24,000
USGS base map had become a standard for planning purposes in Hawaii
and elsewhere (USGS, 1986). Users included the DLNR Parks and
Historic Sites Division (park locations and boundaries), the Water and
Land Division (rainfall and data collection stations), the Land Use
Commission (boundary maps showing urban, rural, agricultural and
conservation districts) and the Department of Agriculture (agricultural
land maps, land evaluation and site assessment).

The following types of information are contained in the USGS 7.5
minute maps: boundaries (such as states, cities, counties, other
municipalities and various administrative entities such as national or state
forests); hydrography (all flowing water, standing water and wetlands);
Public Land Survey System (PLSS) — the rectangular system of land
surveys administered by the US Bureau of Land Management;
transportation (roads/trails, railroads, pipelines and transmission lines);
and other significant man-made structures (USGS, 1986). In addition, in
1985, new data on hypsography (contour data), surface cover (vegetation
such as woods, scrub, orchards and vineyards) and non-vegetative
surface features (i.e., lava, sand, gravel) were added to the USGS list of
standard digital products.

Recently, USGS has begun to offer digital files (Digital Line Graph
and Digital Elevation Model files) containing all the data needed to
reconstruct a computerised version of 7.5 minute quadrangle maps. At
present, USGS has almost completed the process of digitising all of the
quadrangle maps in Hawaii. Hawaii will be among the first states in the
country to have a completed set of digitised 1:24,000 scale maps. An
agreement made between the Office of State Planning and the USGS to
share some of the costs associated with digitising the quadrangle maps has
also helped to put the plans for Hawaii ahead of schedule.

For four of the five major populated islands in Hawaii, the USGS 7.5
minute quadrangle maps provide a ready-made and logical system for
data management. These islands (Oahu, Maui, Kauai and Molokai) are
sufficiently small for it to take only five to seventeen quadrangle maps to
represent an entire island. (See Figure 5.1 which depicts how the
information is organised by quadrangle.) The largest island, Hawaii,

however, requires 74 quadrangle maps to achieve complete coverage. The GIS applications which are emerging out of the state's system primarily involve large-scale resource management efforts. The GIS will be used, for example, to assist in the state's five-year boundary review process of land-use classifications (urban, agricultural, rural and conservation).

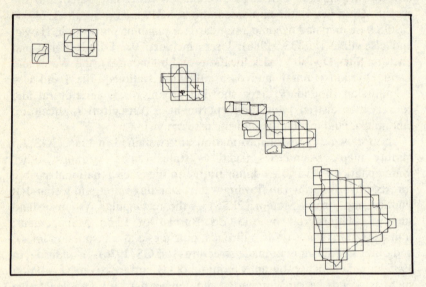

Figure 5.1 *State of Hawaii GIS : State map library tile boundaries*

Another application which has been developed and put to use involves a register of historic sites. In addition to being able to produce site maps with detailed information on important historic and cultural resources, the application also contains a bibliographic reference function. Another application likely to make use of the GIS technology involves the reapportionment of legislative seats which is mandated to occur every ten years as seats in both the State Senate and House of Representatives are based upon population distributions. The primary data source for this application will be the TIGER files of the US Census Bureau.

Another potential application for GIS technology involves project planning. Presently, the state is involved in a number of major development and redevelopment projects which includes the commercial revitalisation of the waterfront in downtown Honolulu, the creation of a new town on Oahu (Kapolei) and a resort-based regional development plan for West Hawaii. These projects have generated tremendous information needs in terms of mapping land-use changes, measuring impacts on infrastructure and public services and planning for the redistribution and allocation of public and private resources.

Efforts to digitise relevant information on native forests, watersheds, isohyets, beaches and shorelines, wetlands, coastal resources and historic sites on the 1:24,000 scale are also under way. While much of the work is being carried out by the OSP and EDPD staff, some of the digitising will be done by various agencies in accordance with their respective areas of interest (for example, the Office of Environmental Quality Control is digitising information on special groundwater protection areas). Other parts of the digitising effort are likely to be contracted out to private sector organisations (e.g. digitising of parks and conservation district sub-zones).

GIS in the City and County of Honolulu

While the state government's perspective on GIS might be characterised as one emphasising resource management, the City and County of Honolulu's perspective is somewhat different. The city's GIS is much more orientated towards permit management, administrative decision-making and assisting in the routine activities of planning, zoning and providing a wide range of vital city services. In spite of the fact that virtually identical technologies have been adopted, the City of Honolulu's GIS is quite different from the GIS developed by the State of Hawaii.

Honolulu's system was developed under a 'turnkey' procurement concept in which the Environmental Systems Research Institute (ESRI) was contracted to design a GIS, to convert and automate various maps and data bases, to develop applications, to install the hardware and software, to provide training and to prepare an organisational plan for the long-term maintenance, operation and expansion of the system. For a fixed sum (approximately US$1.64 million), ESRI agreed to 'turn over' a fully operating GIS which the city would then own and operate. The turnkey approach, which has been used for procuring large, expensive systems (i.e. those covering mass transit systems, sewer lines and various capital projects etc.), helps to minimise cost overruns and to ensure the timely completion and delivery of the procured goods.

As the developer of ARC/INFO, ESRI is one of the most experienced in terms of developing GIS data bases. At the same time, by selecting

ESRI as the consultant to evaluate hardware and software and to recommend a system for Honolulu, the selection of ARC/INFO was an inevitable outcome. The city's system is similar to the state's in that it uses a Prime minicomputer and similar input and output devices. It is currently, however, a smaller system, running on the Prime 4050 (6 megabytes of main memory, 1 gigabyte of disk storage). The cpu and mass storage devices are kept within the Department of Data Systems, and Tektronic 4211 workstations are connected by direct lines. The Prime minicomputer is also directly connected to the city's IBM mainframe system (3090) allowing additional mass storage as well as access to previously developed data sets. In addition, the system can be accessed by IBM PS/2 Model 70 microcomputers which have been upgraded to have Tektronics terminal emulation capabilities. These are connected to the Prime minicomputer either through direct lines or through the city's local area networks. The city plans eventually to eliminate central processing and data storage on the Prime and to develop a distributed data processing system utilising intelligent workstations. All of the equipment will be networked through either existing LANs or a proposed wide area network (WAN). Digitising equipment includes the Calcomp 9100 and the Summagraphics Micrograph II, while output devices include Calcomp 8-pen plotters, laser printers and thermal copiers.

Like the state's system, Honolulu's GIS is a vector-based system and stores tabular data (attributes) in association with simple cartographic features (points, lines and polygons). The mapping data are stored in tables 'standing in relation to other attributes instead of being stored as graphing primitives or symbols' (Dangermond, 1988, p.31) which allows users to view, map, query and to analyse spatial information. Unlike the state's GIS system, however, which is built on the USGS 1:24,000 scale quadrangle maps, a new collection of base maps using the city 1:200 and 1:400 scale zoning maps has been developed for the city GIS.

The base map data bases contain information which is used for geodetic control (vertical and horizontal) and spatial referencing as well as data on major public facilities including buildings, airports, parks, historic and cultural sites. Additional layers consist of lot and parcel boundaries, building or address centroids, easements and other land record data. The GIS also includes a transportation data component in which the street network, rights-of-way, bus routes, traffic signals and other transportation data have been geocoded. The GIS data base includes layers with extensive information about water, sewer and storm drainage systems. Finally, the GIS contains information about the various boundaries systems and areas used in public administration: these include legal (land-use, zoning and special districts), political (City Council,

Neighbourhood Board, Legislative) and service planning (police beats, refuse collection districts and sewer improvement districts) boundaries.

With GIS systems it is often convenient to think of the information being kept in two dimensions: while 'layers' or 'coverages' represent the overlays which contain different types of information for a specific geographic area, 'tiles' represent the way a given map or layer is broken into smaller, manageable pieces. The tiles form a referencing system enabling the user to call up only the needed portion of a particular map. Accordingly, the tiling system says much about how the system can be used and what level of information is contained in the data base. Herein lies one of the major distinctions between the city and state systems in Hawaii. While the state's tiles are defined according to the USGS (1:24,000) quadrangle map boundaries, the city's tiles are based on smaller units with each 7.5 minute quadrangle map having been broken down into nine district areas of equal size (2.5 minutes by 2.5 minutes). Each of these 2.5 minute quadrangles was further broken down into four equal sized tiles which were 1/36th the size of the 7.5 minute quadrangles. Each tile corresponds to a geographic area of just over one square mile. It is important to note, however, that the tiling system can be altered: for example, ARC/INFO has the ability to recognise non-rectangular tiles so that spatial data can be organised in terms of development plan areas or other logical groupings.

The smaller tiles in the city's system allow the storage of much more detailed and intricate data at a more accurate scale than the state's system. The parcel layer depicts polygons which are legal lot boundaries and contains information on lot dimensions and site addresses (see Figure 3.2 which illustrates a parcel map with water lines). Over 140,000 parcels have been digitised. By interfacing the GIS with the city's other computerised administrative files — including the Real Property Tax Administration data base in the Department of Finance and the Land Data Base (which contains information on land use permits and general planning) — the amount of available parcel-specific information can be greatly augmented.

Municipal government applications

To understand the GIS applications which are emerging in Honolulu, it is important first to understand a little about the City and County of Honolulu. The County has jurisdiction over the entire island of Oahu. It is the largest in population of the four counties in Hawaii and has had a tradition of strong municipal management. The Department of Data Systems manages an IBM 3090 mainframe computer and provides

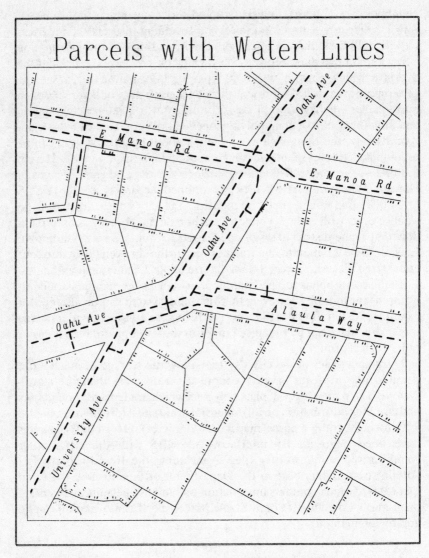

Figure 5.2 *City and County of Honolulu: parcels within water lines*

computing services not only to the various city departments but also to the other counties as well. Even before implementation of GIS, a number of city departments were already making use of computerised technologies to aid in planning, in the administration of the permit system and in managing municipal services. The director of the Department of General Planning (DGP) has been one of the driving forces behind the development of Honolulu's GIS. He has acted as both catalyst and coordinator to ensure broad participation and integration of the GIS with existing systems. The involvement of senior management has been important to the general success of implementing GIS within municipal management in Honolulu.

Honolulu, which is unusual in the sense that the planning and zoning functions have been separated into two departments, also has a Department of Land Utilisation (DLU) which is responsible for administering the zoning code. One of DLU's primary responsibilities has been the preparation and updating of the city's zoning, land-use, flood zone and shoreline maps for the GIS. Another department providing substantial input into the development of the GIS is the Department of Finance (DOF) which is responsible for the preparation of tax maps showing parcel boundaries. The GIS will enable, for the first time, the production of computerised maps containing information derived from the three departments of General Planning, Land Utilisation and Finance.

The Honolulu GIS will assist in some aspects of broad policy planning. For example, information is needed to evaluate proposed changes in the General Plan as well as in various area-specific development plans. The GIS will also provide information for an Industrial Land Needs Study and a Growth Management Policy Review as well as more general studies which require data on population, housing, development capacity and general policy issues. For instance, the Honolulu General Plan calls for controlled growth of the island's population to 'avoid social, economic and environmental disruptions'. The Plan establishes target level residential population distributions for the year 2010 by location and the GIS will enable the more periodic and up-to-date assessment of how well the objectives of the General Plan are being achieved. Another objective involves maintaining the viability of Oahu's tourism industry. The GIS will provide location-specific information regarding Oahu's visitor plant industry and assist policy-makers in identifying critical areas for public investment and infrastructural development. GIS, thereby, provides a tool for long-range planning.

While Honolulu's GIS can be utilised to assist in long-range planning and resource management (similar to the state's system), its strength lies in the applications involving management of permits and administrative decision-making. Land development in Hawaii is a highly regulated activity. In addition to federal permits (Corps of Engineers,

Environmental Protection Agency, Federal Aviation Administration etc.), there is also a virtual jungle of state (Department of Health, Land and Natural Resources, Environmental Quality Control, Transportation etc.) and local (Board of Water Supply, Building Department, Fire Department, Planning, Land Utilisation, Parks and Public Works) permits and approvals which must be cleared before development can be initiated.

A GIS system can be especially useful in identifying the circumstances under which approvals or permits must be sought. The GIS can assist in terms of identifying jurisdictions or districts in which special controls or regulations apply. Developments which occur, for instance, in coastal areas, in conservation districts, in flood zones, in airport hazard zones, in special management zones or in special design districts are all subject to different development restrictions. The GIS can be used to pinpoint, either by parcel or perhaps by development plan area, the various restrictions and regulations which may apply to development. This is vital information for both the prospective developer or citizen interested in a specific parcel, as well as for the public servant. The planner or public official must often decide whether or not a certain permit is required and in some cases make an administrative ruling (e.g. requiring a conditional use permit) or prepare information for a review body (e.g. the land-use commission or the zoning board of appeals). Because the city's GIS will have information at the parcel level, the GIS can be used both to answer queries and to aid decision-making.

Zoning and land-use regulation is often said to involve the use of the government's 'policing' power. The city GIS can also be used to assist in evaluating compliance with various land-use regulations (for example conformity of uses, set-backs, height and bulk restrictions). One problem area in Hawaii is the 40-foot ocean shoreline set-back (this is a zone of special land-use and development control). Because of the demand for ocean-front property, some owners have constructed illegal structures in the set-back area. Others, in an effort to protect their properties from coastal erosion, have built illegal sea-walls or revetments. Because of lax enforcement, there is a perception in some areas that such illegal practices will be tolerated by government. As a result, in these areas there has been widespread violation of the shoreline set-back rules. A GIS will help to monitor and track compliance in 'trouble spots' as well as assist in efforts to bring properties into compliance.

Honolulu's GIS is also designed to assist various departments in several aspects of public administration. For example, within the public works department, the GIS will help with impact analysis of the city's trash incinerators, transfer stations and landfill facilities. Other public works applications include the scheduling of road resurfacing, street sweeping and the maintenance of the sewer system. A variety of applications have been developed for the Honolulu Police Department: these include the geographic analysis of crime, assistance with the deployment of officers and vehicles by beat, sector and district, the analysis of road traffic accidents and emergency/disaster planning.

One of the principal differences between the city GIS and the state GIS is the extent to which the city's system is built around serving routine applications generally involving maps and spatially-oriented data. The state's system is much less task-oriented and very much less geared to the immediate needs of a particular agency or department. The strength of the state GIS lies in its coverage of important environmental data and its extensive coverage of natural resources. Taken together, the two systems complement each other and provide a logical division of labour. Whereas the 1:24,000 scale is quite appropriate for many state-level decisions, the day-to-day operations of municipal government require a much finer resolution and are much more accuracy-dependent.

The future of GIS in Hawaii

The future of GIS in Hawaii looks full of potential. The initial large investments in hardware, software and data bases by the State and City and County of Honolulu suggest that Hawaii is one of the prime movers in the field of applying GIS in a practical planning and management environment. Because both the state and local government have chosen similar systems and have used the same software vendor, many of the problems associated with system compatibility have been minimised. It should be noted that the USGS Water Resources Division in Honolulu also uses ARC/INFO operating on similar hardware. In Hawaii, there is great potential for data sharing and for using GIS to improve administration and decision-making at various levels of government. There will also be a core of ARC/INFO users in the state. While it is still too early to conduct a full evaluation of either GIS, certain basic aspects of the technology have emerged in Hawaii which warrant further discussion. The future of GIS in Hawaii is inexorably linked to the following concerns: the quality of data; access to data; the maintenance of data; training and education; the development of new applications; and the continued interest in, and development of, spatially-oriented planning.

On data quality, the expression 'garbage in, garbage out' is especially relevant to GIS systems. Very strict standards must be applied to the digitising of maps as well as the inputting of data to attribute tables. Because of the vast amounts of information being stored, there is the problem that errors can go undetected for long periods of time and may even propagate within the system. Some data being entered in at the parcel level, for example, may not be used, if for instance, the area is one in which little or no growth is occurring. There is also the problem that people are much more inclined to believe that computer-generated maps and print-outs are more accurate and credible than those produced by hand: this is not always the case. Often, certain short-cuts in data acquisition may have been taken and it is difficult to ensure common standards of data quality if that data has come from many different sources: hence, the quality of the data must be tested before it is included in the system. A particular problem of data quality arises from the assumption that the USGS (1:24,000) scale maps were accurate before they were digitised. A variety of different types of errors has been found in USGS maps (Carter, 1989, p.255) and efforts to improve these data bases continue. There is also an overriding need to set up more in-state protocols for data testing and quality control. Currently, discussions on developing standards for GIS data, such as those proposed nationally, are under way in Hawaii (see *American Cartographer,* January 1988).

A second major concern affecting the future of GIS involves access to data. Already issues regarding the proprietary nature of data contained in the GIS have emerged. It is easy to understand why: the costs of digitising information are significant and it is only natural that those bearing these costs should be concerned about access to, and the use of, data. In part, these costs explain why government, rather than the private sector, has taken the lead role in the development of GIS data bases in Hawaii. While there are certainly private sector organisations such as real estate companies, utilities and information services who are interested in gaining access to GIS data bases, policies regarding the transfer of data have yet to be worked out and ways of cost-recovery or cost-sharing have yet to be considered. With a new technology such as GIS, the front-end costs tend to be large and often government has subsidised its development and expansion. Public access to data may also involve issues of confidentiality, although most of the data being automated in Honolulu is either publicly available or concerns environmental, legislative or other other public interest concerns.

Even within the public sector itself, agreements as to how and when the State and County should begin sharing data have yet to be agreed even though there are numerous precedents for data sharing agreements between levels of government in Hawaii. This is a positive aspect of GIS technology in that it presents a new opportunity to set aside partisan

politics and to establish ongoing communication and cooperation
between the State and County governments in Hawaii. At present, the
Governor and most of the State Legislature are Democrats, while three
of the four County governments have Republican mayors.

Access across departments within individual agencies is yet another
concern. While interdepartmental access to a centralised data base is
beneficial from the point of view of data sharing, it creates data
management problems and increases the potential for unauthorised use
or perhaps damage (intentional and unintentional) to the data base itself,
even though routines in a system such as ARC/INFO can help minimise
this problem. Many of these problems can be controlled through sound
data base management. It cannot be doubted that there are advantages to
be gained from adopting the minicomputer versions of the GIS software
in that data files tend to be more centralised, the hardware requires a
system manager and support personnel and there is greater control over
users than if microcomputer or stand-alone systems were used.

Another major concern which has already emerged in Hawaii, even
though the data bases themselves have not yet been fully completed,
involves the issue of maintaining and expanding the data bases. The
reality is that Hawaii is changing so quickly and without a long-term plan
for upgrading the system it will rapidly become out-dated, less reliable
and less likely to be used. As more agricultural land is converted to urban
uses, as new infrastructure is built and as more homes are developed, the
GIS must keep pace. While it may be easier to track new subdivision
developments than infill developments or the redevelopment of existing
properties, both types of change need to be routinely entered into the
GIS if it is to be an effective planning and management tool. Perhaps
developers of new subdivisions should be required to pay the additional
digitising costs of having their projects included in the GIS data base. This
would help to defray costs and also to ensure that vital new information
for large areas is included as it becomes available.

Central to the future of GIS in Hawaii is the training and education of
system programmers and users. There is but a handful of GIS
programmers in the state and an equally small number of ARC/INFO
users. The University of Hawaii is the major source of planners and
geographers in the state and efforts are under way in Hawaii and
elsewhere to increase the number of University courses offering GIS
education and to expand the volume of research in GIS applications. Two
problems have arisen: first, GIS technology is currently very expensive,
and second, ARC/INFO is not an easy system to learn. This potentially
makes the gap between the academic-theoretician and the professional-
practitioner wide, but as costs decrease and more and more support for

GIS evolves within the state, the gap will narrow. It is not unreasonable to expect that within the next few years, GIS will become a standard component of the planning and public administration education in Hawaii.

Within the sphere of planning, there is undoubtably a need to develop more applications using GIS technology and methods. GIS is a relatively new technology for planners (Levine and Landis, 1989, p.209) and because the technology is so new, there is still room to develop new and powerful applications. However, there is much confusion among the planning profession particularly over the definition of GIS: some planners still have trouble in distinguishing between GIS, Computer Aided Design (CAD) systems and Computer Aided Mapping (CAM) systems (Dueker, 1987, pp.383-4). In addition, a major reason why more applications have not been developed is that planners generally confine themselves to a personal computer (PC) environment: this may unnecessarily limit their horizons as, in the majority of offices, the use of PCs is not particularly sophisticated. The evolution of PC-based GIS technologies will undoubtedly help to increase the overall use and development of GIS but may impose constraints on the size and complexity of applications. At present, GIS technology is treated largely as a tool to aid in decision-making and applications almost invariably involve only the presentation of information for decision-makers: the potential of GIS systems goes well beyond this simple presentational function, and as the technology becomes more accessible, increasingly diverse and sophisticated applications are likely to emerge. For instance, GIS technology might also be used in public meetings to assist in resolving the conflict which often arises over land-use proposals or in displaying the implications of development proposals derived either from 'scenario writing' or from more rigorous mathematical simulations.

To a large degree, the future of GIS in Hawaii and elsewhere is largely dependent upon a continued interest in spatially-oriented planning. The allure of GIS lies in its capacity to represent reality as either a two-dimensional or three-dimensional map or drawing and to illustrate graphically future changes which result from plans and policies. It is only natural that in Hawaii, where land and space are so precious, GIS technology has been so well accepted. As the limits to growth become apparent in other places, so too will the attractiveness of tools such as GIS.

Acknowledgement

I would like to acknowledge the following people for the comments they have made on earlier drafts of this paper: they are Dr Norman Okamura and Joan Esposa (both of the Department of Budget and Finance, State of Hawaii) and Donald Clegg and Gary Hamilton (both of the Department of General Planning, City and County of Honolulu).

Bibliography

American Cartographer (1988), 'Proposed standard for digital cartographic data', 15(1) January.

Armstrong, R. (1983), *The atlas of Hawaii,* (2nd Edn), University of Hawaii Press, Honolulu.

Carter, J. (1989), 'Relative errors identified in USGS gridded DEMs', *Auto-carto 9,* Proceedings of the Ninth International Symposium on Computer-assisted Geography, American Society for Photogrammetry and Remote Sensing and American Congress on Surveying and Mapping, Falls Church, VA.

Dangermond, J. (1988), 'GIS wins', *Planning,* 54(12), pp.31-2.

Department of Business and Economic Development (1989), *Hawaii State data book,* Department of Business and Economic Development, State of Hawaii, Honolulu.

Dueker, K. (1987), 'Geographic information systems and computer-aided mapping', *American Planning Association Journal,* Summer, pp.383-90.

Gwynn, W. and Ogorski, C. (1987), 'Design and technological aspects of U.S. Geological Survey maps', *American Cartographer,* 14(3), pp.203-14.

Jannace, R. and Ogorski, C. (1987), 'Cartographic programs and products of the U.S. Geological Survey', *American Cartographer,* 14(3), pp.197-202.

Levine, J. and Landis, J. (1989), 'Geographic information systems for local planning', *American Planning Association Journal,* Spring, pp.209-20.

Lima, R. (1984), 'Planning software survey', *American Planning Association',* Planning Advisory Service Report no. 388.

Maguire, D. (1989), *Computers in geography,* John Wiley and Sons, New York.

Mueller, J.C. (1985), 'Geographic information systems: a unifying force for geography', *The Operational Geographer*, 8, pp.41-43.

Office of State Planning (1988), *State automated geographic information systems'* a report to the 1988 legislature, Office of State Planning, Office of the Governor, State of Hawaii, Honolulu.

United States Department of the Interior, US Geological Survey, (USGS) (1986), 'Digital line graphs from 1:24000 scale maps', *Data users guide*, Reston, Virginia.

Chapter Six
Developing Geographical Information Systems in local government in the UK: case studies from Birmingham City Council and Strathclyde Regional Council

Iain Gault and David Peutherer

Introduction

Within this chapter we propose to examine the development of geographic information systems in UK local government from a different perspective from the ones which are generally used in GIS case studies. Many of the evaluations of the potential use of GIS in practical policy-analytical situations begin either from a technological starting-point or from an information handling starting-point. In this paper, we propose to examine the development of GIS in the context of the broader political, managerial and organisational trends and forces currently affecting UK local government. For, as Smith (1989, p.54) pointed out, 'the development of effective information systems is more often hindered by institutional and managerial factors than by technical ones'. In our discussion, we will focus relatively little attention on the technological environment in which GIS is being developed as this is covered elsewhere. Rather, we will place most attention on examining those changes within UK local government which are forcing decision-makers (both political and managerial) to make far more effective use of the information in their possession, which in turn is forcing them to develop more sophisticated analytical techniques to improve the quality and timeliness of the decisions they have to make.

In this chapter, we will paint a picture of the changing environment affecting UK local government and examine the impact of both externally determined factors (such as recent legislative change and the changing financial climate) and internally determined factors (such as the increasing local political desire to target resources more effectively on special needs groups and disadvantaged communities and the increasing role of consumerism in the planning and provision of services). We will then examine various examples of current practice in local government in an attempt to identify the impact that these factors have had on the development of (usually medium-term) planning frameworks, on the clearer articulation of information needs, on the development of better statistical systems and, more recently, on the development of GIS.

Our two main examples will be the development of a pilot GIS study in the City of Birmingham and the development of policy-orientated statistical systems and planning frameworks in the Strathclyde Region of Scotland. Throughout our review, we will also refer to supporting examples drawn from elsewhere within UK local government. After examining this contextual material, we will then examine some of the issues relating to both the development of information strategies within local government and examine some of the problems which must be overcome if GIS are to become more widely used. We will also discuss some of the lessons we have learned about the development of more effective and needs-sensitive medium-term planning frameworks, about the development of local statistical systems and about the organisational implications of GIS development and use.

The changing political, managerial and organisation environment in UK local government

In recent years, the environment within which local government in the UK functions has changed significantly. In the foreseeable future, it is unlikely that many of these changes will be reversed and, indeed, it is more likely that the process of change will continue in the same direction. The changes affecting UK local government can be divided into those which are being imposed upon local government by central government and into those which are being generated from within local government. These changes are not independent of each other — rather, there is a significant interaction effect between them. The impact of these changes has already resulted in some authorities adopting more structured and rational decision-making procedures, and, under the pressure of the changes taking place, more authorities are likely to follow suit, either because they themselves recognise the value of so doing, or because they are required to do so by central government. It is important to consider the potential role of GIS in this context.

Currently local authorities in the UK are having to cope with a large volume of legislative change imposed upon them by central government. The areas of change include a fundamental change in the method of local taxation (away from a local taxation system based on property towards a personally based 'Poll Tax' system), a change in the provision of education locally (Education Reform Act, 1988), a change in the way that personal social services are provided (Griffiths, 1988), changes in the way that housing is both provided and managed (see Clapham, 1989) and also changes in the way that local government has to conduct its business (HMSO, 1986).

In addition to these structural changes, local government in the UK has been put under substantial financial pressure throughout the 1980s (Flynn, 1989, p.99). Indeed, the prevailing trend within UK local government is towards the creation of what has been termed 'enabling' local authorities, whose role is not the direct provision of services; rather its role is seen as providing a strategic management framework in which services can be provided either exclusively by the private sector or by the public sector in direct competition with the private sector. If these trends persist, the morphology and role of local government in the UK will change dramatically in the 1990s.

The implicit danger of this comprehensive restructuring of the role and operating environment of local government is, that by having to concentrate their resources on responding to these legislative requirements, local authorities will fail to recognise the underlying shift in the nature and purpose of local government, and will fail to put in place mechanisms which are expressly aimed at the controlled management of change — GIS is one of those mechanisms. Our main arguments are first, that they need to adopt more robust and rational decision-making procedures than they have at the moment if they are to manage change more effectively and second, that if local authorities are to manage change in a positive and locally responsive way, then they will need to develop new information sources, make more coordinated use of the information currently at their disposal and develop more effective ways of handling, analysing and presenting that information.

In addition to the pressures emanating from central government, stimuli for change are coming from a variety of internal and external sources: these include elected members, senior managers, local communities, the private sector and other public agencies. These radical changes have had a fundamental impact on financing methods, on administrative procedures and on priority setting. These changing pressures have challenged the very basis of local government as it has been practised since the reorganisation of the early 1970s (HMSO, 1972).

The impact of these changes has meant that, since the late 1970s, many authorities have, largely due to central government financial restrictions, either had no growth, or suffered reductions in the real value of their revenue and capital resources. The introduction of the Community Charge or Poll Tax (in Scotland in 1989 and in England and Wales in 1990) has imposed further constraints on expenditure in that any increase over and above that funded by central government will need to be funded locally through the Poll Tax. The implications of this are that the full financial burden of developing services locally will fall directly on local residents (John, 1989, p.4). Elected members will come under pressure locally to avoid increases which could be viewed as excessive by the electorate or which might lead central government to restrict the ability of authorities to set their own Poll Tax levels.

Although many authorities have responded by voluntarily taking measures to improve their operating efficiency, changes are also being imposed by central government and bodies such as the Audit Commission to force local authorities to increase their effective use of resources, to carry out value-for-money studies, to set objectives and targets for services and for local authorities to monitor more rigorously their spending and performance against these objectives and targets. For some, these developments constitute a wholly new concept of management, requiring the introduction of radically different procedures for the more effective management of physical resources and the more responsive allocation and targeting of policies and programmes.

An additional impact of central government imposed legislative change and the enforced move towards local government having an enabling role rather than a direct provision role, is that local government has become more overtly political. One impact is that elected members in many authorities have become more interested in determining policies, programmes and priorities locally, and in monitoring the effectiveness of service delivery and resource utilisation in relation to these policies. A good example of this is the Strathclyde approach to the development and implementation of social policy of which an integral component was the designation of 'Areas for Priority Treatment' which were based upon localities with a community identity within which there were concentrations of multiple deprivation. The Strathclyde approach is described in more detail in the documents *A social strategy for the eighties* (SRC, 1980), *Strathclyde — generating change* (SRC, 1988) and in Young (1989).

Over the same period the electorate has become more demanding and there has been a considerable growth in 'consumerism' in the planning and delivery of local services. Elected members have consequently become more sensitive to the needs and aspirations of their constituents, and this has led to an increased interest in assessing consumer needs and preferences which is reflected in the greater use of consumer surveys by local authorities (Hancox *et al.*, 1988). Many local authorities are now developing marketing strategies, which, as in marketing in the private sector, are largely concerned with identifying target groups and the communities in which they live as a basis for the more effective marketing and targeting of services. GIS has played a major role in marketing in the private sector (see Beaumont, 1989).

Several local authorities (for example, the London Borough of Islington, Walsall Metropolitan Borough Council and Wrekin Council) have decided to decentralise the way in which they deliver and plan services, and this is in direct contravention to the centralisation which took place at the 1974 reorganisation of local government. In those local authorities which have decentralised, services are increasingly being planned to meet community-determined goals and being managed at the local community level with greater participation from within the communities themselves. The impact of this trend towards greater decentralisation is that information flows have become more diffuse and that there is now an overriding demand for information to be provided at the community level to a non-technical audience (Pay, 1989).

There has also been an improvement in the quality of management within local government. This can be traced back to the mid-1970s with the 1974 reorganisation of local government and the introduction of corporate planning and management, but more recently it is a reflection of the increased investment in management training which has created a generation of younger managers more receptive to the use of new management techniques. This has led some authorities to adopt a strategic medium-term planning approach which provides the overall policy framework for the development of services and expenditure plans and for influencing, via inter-agency strategies, the planning activities of other organisations (Worrall, 1989).

A key characteristic of this more formal, scientific and systematic approach is that it requires the use of more rational decision-making procedures which incorporate the monitoring and forecasting of conditions, needs and opportunities; the identification and prioritisation of issues; the setting of clear objectives and targets; the analysis of options; the formulation of policies and programmes; the monitoring and evaluation of performance against objectives, and the regular review of the authority's strategy, policies and programmes. As local authorities are legally required to produce an annual budget, it is essential that much of the information in this planning cycle is refreshed at least annually. The changes that have affected, or are about to affect, local government have increased the need to adopt a more systematic approach to management. The existence of better trained managers and the availability of new technology are important elements, but the key to the implementation of these procedures is the availability and management of information and the availability of tools like GIS to facilitate its intergration, analysis and presentation.

A review of local government information needs

The successful implementation of the main elements of the rational planning approach set out above depends on the availability and appropriate use of accurate information. For example, if local authorities are to identify and prioritise issues, they need to monitor and forecast the changes affecting their areas and services. This can require a wide variety of information relating both to the number and situational characteristics of their clients and to the resources available (these include personnel, finance, land, buildings, roads, etc). In addition, information on changing infrastructural, social, demographic and economic conditions is essential.

Local authorities exist primarily to provide services for people and so they need to be able to formulate relevant policies to meet the needs of certain client groups: in particular, they need information about the numbers in various groups and likely compositional trends (e.g. the unemployed or the elderly), information about needs and aspirations and how these might change in the future, intelligence about the success or otherwise of past attempts to assist that particular group and information about the policy options available. Local authorities also need to set realistic targets for future service provision: hence, information is needed not only about the size of the client group but also on the implications of providing services to varying standards and on the likely availability of finance and labour.

Local authorities are major landowners and can thus have a major impact on the physical development of areas: for example, the City of Birmingham has an extensive land portfolio and is the largest landowner in the City of Birmingham. Local authorities are also responsible for the maintenance and repair of highways, for public open space and are usually the predominant landlord in individual localities: for example, in 1988, there were over 465,000 rateable hereditaments in the City of Birmingham and over 2,100 kilometres of adopted road (Gault and Davis, 1988). In the financial year 1989/90, Strathclyde Regional

Council, with its 96,000 employees, had a turnover of £2 billion (Peutherer, 1988).

We have established that local government is a major industry in its own right and is responsible for providing a wide range of very different types of products and services. These services include personal services such as social services, education and housing; protective services such as police, fire, public health; and infrastructure services such as roads. Local government is responsible for both strategic planning (including transportation) and detailed land-use planning and development control. All these activities have to be planned and controlled not only on a day-to-day basis, but also with regard to the future use and deployment of resources between areas, groups, services and products.

Taking account of this diversity of activity, and the need to plan, it is possible to identify certain main information requirements. Local authorities need many different types of information: for example, they need regular information relating to social, demographic, economic and land-use conditions; the availability of resources; and the needs and aspirations of the community. Most local authority activities are carried out on varying geographical units. The units include service delivery areas (e.g. school catchment areas), policy and programme areas (e.g. inner-city areas) and community areas (e.g. neighbourhood office areas). Unfortunately, most of these specialised spatial units are not coincident and their size varies considerably. Information is needed about each area in a form which meets the needs of particular service managers, groups of service managers or the elected representatives involved. To meet these needs, information has to be available at a small area level, to be capable of flexible use and capable of being integrated with other data sources. The data also has to be capable of aggregation to any area for which it may be required.

Because of the complex structure of local government in the UK, information is collected by a wide variety of agents for many different purposes. (There are two tiers of local government throughout most of England, Wales and Scotland — except in the metropolitan areas of England where there is only one tier, the second tier having been abolished in 1986. Health Authorities and Water Authorities are entirely separate from local government in England and Wales though Water and Sewerage is a local government function in Scotland.) Information collected by local government is often exported to other organisations for their use. Much of the information required by authorities is collected initially for administrative, managerial or control purposes relating to individuals, utilities or facilities, though much of this information is subsequently used for planning purposes: for example, in assessing the overall need for, and the effectiveness of, particular services.

However, these 'value-added' uses of the information can only be realised if it is possible to cross-relate and integrate diverse data sets and this implies the need for effective cooperation both between separate departments within local authorities and between public bodies: more concisely, there is a need to create information synergy and it is here that GIS can make a major contribution. For example, while local authorities tend to be the custodians of small area population data, liaison is needed with the police authorities to ensure that the way that crime data is collected and stored is compatible with how that population data is structured: the subsequent relating of population data to information about the incidence of crime allows the generation of crime rates, which can provide an input into identifying priority areas for community-based policing initiatives.

Because of the multi-functional role of local authorities, and because of the need to use information for operational (administrative functions), managerial (control functions) and strategic (planning-related) purposes, the issues which surround information collection and use in local authorities are very complex. In some cases, as with population data, there are many potential users and uses within an authority and the information held has to be capable of being delivered to these various users in the right form and at the right time to meet their particular needs. At these different functional levels within local authorities (operational, managerial and strategic) there is a need to be able to relate past, current and projected information for the effective planning and delivery of services to be achieved.

There is also a need for both 'hard' and 'soft' information (with the latter particularly comprising data on community attitudes) though the need for soft (attitudinal and perceptual) information is highest at the top of the organisational pyramid (i.e. at the strategic planning level). As local authorities increasingly decentralise their services, this soft information will increasingly be needed at lower levels within the organisation and at a finer spatial scale. The ability of GIS to handle both hard and soft information in a systematic way will thus become one of its critical success factors.

Information management issues

The combination of the nature of the changes facing local government, and the characteristics of the information required to manage these changes at different levels within the organisation, leads to certain inescapable conclusions as to the way in which this information should be managed and organised. In Table 6.1, we list some of the objectives that information systems in local government should fulfil if they are to enable the effective management of change.

Table 6.1 Information systems design objectives for local government

1. Information systems should be designed corporately to ensure that the needs of all management levels and parts of the organisation are met: the key groups with specific and distinctive information needs are:

   a. Senior management and strategic planners
   b. Service managers and function planners
   c. Service deliverers
   d. Elected representatives
   e. The public
   f. 'Issue-based' or community-based special interest groups

2. The systems should be able to provide each part of the organisation with the information required, when it is required, but avoiding both information overload and information starvation.

3. Information has to be available to be used selectively and used to measure progress towards planning objectives.

4. The systems should be integrated, interactive and networked to allow mobility of data between sources and users.

5. The systems should be able to provide information on a small area basis but should also have the flexibility to produce information about the whole range of spatial units (i.e. school catchment areas, 'communities' or programme areas) which are of interest to the organisation.

6. The systems must be capable of holding or generating, past, current information to allow time-series analysis. The systems should also contain forecasting or simulation modules to enable trend extrapolation and policy testing.

7. The data held must be of good, or at least defined, quality. Good system design is no substitute for good quality data.

Our analysis of the nature of the pressures facing local government, and our subsequent discussion of the information system design features that are required to assist in the management of change, gives rise to three main implications for information management in local government. First, because of the multi-functional nature of local government, information systems must be capable of being integrated as policy development will require that actions across a broad range of functional areas are coordinated often within spatially definable communities. Second, the analysis of information must concentrate on the complex and changing relationships between locally determined needs or priorities and the resources available — if services are to be provided in a way that is acceptable to client groups, significantly more effort must be applied to demand forecasting and assessment than is done at present (Derbyshire, 1983). And, third, information must be capable of being displayed or presented in a manner which aids the understanding of spatial distributions and spatial patterns at a variety of (geographic) levels and for a wide variety of users (many of whom are neither 'spatially literate' nor numerate).

We have defined the changing operational, political and managerial context within UK local government: the prevailing implications are essentially that local government must develop the procedures and information systems to enable it to plan its services more responsively and to manage its resources more efficiently. We will now examine, first, how information systems have been developed and coordinated, using Strathclyde Regional Council as an example, before turning to an examination of the application of GIS to the development of more effective and efficient management methods, using Birmingham City Council as an example.

Information systems development in Strathclyde Regional Council

Throughout the 1980s, Strathclyde Regional Council has adopted an interventionist approach towards social policy which has been orientated towards improving the life chances of those who are least able to compete in the labour market, the housing market and other systems which determine the allocation of society's resources. The first step in the process involved the identification of those parts of the region experiencing grave and particular social, economic and environmental problems as a basis for assisting the Council to deploy its resources more sensitively. Given our view that it is essential to have high quality, regularly updated and accurate information to serve the strategic planning process, Strathclyde Regional Council has applied substantial resources to developing information systems capable of supporting its strategic objectives and policies. Thus, there is a very strong demand within the Council for regularly updated small area information which is directly relevant to the monitoring and evaluation of the effectiveness of Council policies and programmes.

The information systems which have been developed can be categorised into three distinct groups: corporate data bases, departmental (or operational) data bases and external data bases.

Table 6.2 Overview of the Strathclyde Regional Council small area
information system

Data source	Basic level of availability	Spatial levels of presentation[1]
Voluntary population survey (Annual)	Unit postcode/ Enumeration District	a, b, c, d, e, g
Unemployment data (Quarterly)	Postcode sector	c, (d, e, g only if aggregations of c)
Vital statistics (Annual)	Postcode sector[2]	c, (d, e, g only if aggregations of c)
Social Services Department records	Unit postcode/ Enumeration District	a, b, c, d, e, g
Education Department records	Unit postcode/ Enumeration District	a, b, c, d, e, g
Finance Department records (from 1990)	Unit postcode/ Enumeration District	a, b, c, d, e, g
Decennial Census[3]	Unit postcode/ Enumeration District	a, b, c, d, e, g
Utilities	O.S. grid reference	a, b, c, d, e, f, g
Land and Property Register	O.S. grid reference	a, b, c, d, e, f, g

Note: 1. spatial levels - (a) unit postcode   (b) Enumeration District
                                     (c) postcode sector   (d) Policy monitoring
                                                              areas
                                     (e) Programme areas   (f) Point/parcel
                                     (g)Ad hoc areas

2. some limited information on vital statistics is available at unit postcode level

3. the data likely to be available from the 1991 Census at the unit postcode level is not yet known

Source: Peutherer, 1988

Clearly, the use of information from such disparate sources has meant that considerable attention has been focused on improving the integratability of the various data bases. This is largely being achieved spatially by the use of postcodes. An overview of the information available from the systems is shown in Table 6.2.

Corporate data bases

Perhaps the best example of a corporate data base in Strathclyde is the local Voluntary Population Survey (VPS) which is described in Black (1985). The face-to-face survey is undertaken annually in conjunction with the electoral registration process: in 1988, there were approximately 800,000 households in the region accounting for a population of 2.3 million. At the lowest level of spatial aggregation, data is available on each of 75,000 unit postcode areas (these contain an average of twelve households). Though this file is potentially available for use, the data has been spatially reformatted into Census enumeration districts (EDs) of which there were 7,450 in the Region in 1981 (each containing 123 households on average). The use of EDs has meant that it has been possible to roll forward the decennial Census on an annual basis: in addition, the ED is a basic spatial data unit which is both small enough to be useful for a variety of planning purposes and large enough to preserve confidentiality.

A summary of the information derived from the Strathclyde VPS is contained in Peutherer (1988) but the usefulness of the VPS extends well beyond the range of indicators which are directly derivable from it. For example, information on one-parent families drawn from the VPS can be used in conjunction with other information as indicators of the incidence of other policy-relevant social issues such as low income or poor housing conditions; the information can be used to assess specific client group needs and can be used as a basis for calculating and comparing small area occurrence rates for crime, unemployment, mortality and fertility — all of which are essential inputs to the formulation and monitoring of community-based social policy.

Departmental data bases

Individual departments within the Regional Council, particularly the Education, Social Services and Finance Departments, compile massive client-orientated data bases as a part of their routine administration of services. In order to obtain full corporate benefit from these systems, a commitment has been made to apply unit postcodes to each client record in each of these three departments. The departmental information systems will then be capable of integration with the VPS at the unit postcode level.

The data systems in the Education Department will furnish information on the uptake of free school meals, on educational attainment and on participation in education after minimum school-leaving age. The unit postcoding of Social Services departmental records will provide information on the incidence of children in care or under local authority supervision, on vulnerable elderly people and on the physically and mentally disabled. Finally, from 1990, the Finance Department expect to hold information on all persons receiving a Community Charge rebate which will act as a proxy measure for poverty when used in conjunction with other indicators drawn from other sources. The VPS will provide the denominators for the calculation of small area social indicators from these operational systems.

External data bases

In addition to the information generated from within the Regional Council, it is possible to merge information drawn primarily from central government departments though not at such a fine spatial scale as the unit postcode. Unemployment data, comprising counts of the registered unemployed by age, sex and duration is available nationally at the postcode sector level (there are 410 postcode sectors in the region). The General Register Office (Scotland) also provides data on fertility (number of births and age of mother) and mortality (deaths by age, sex and cause), again for postcode sectors. The VPS is used to provide the complementary information necessary to produce standardised mortality rates and standardised fertility rates at this spatial scale.

The development of automated cartography and the evolution towards GIS

In 1986, a Computer Based Mapping Group was established within the Regional Council to examine the feasibility, cost and potential uses of GIS and to determine how progress might be made towards its development on a corporate basis. After a pilot exercise had been undertaken, it was concluded that the introduction of a GIS would be of major benefit to the Regional Council for the following reasons: it would provide greater ease and speed of collecting, updating and analysis information which would culminate in more effective information use; it would improve the accuracy, currency and comprehensiveness of the information available; it would provide the medium for linking individual data sets; it would enhance the interchangeability of information between different users; it would necessitate the development of common data standards across a very large organisation; and it would enable the Council to pursue its social policies more effectively and manage its resources more efficiently.

Within an organisation as large as Strathclyde Regional Council, the time needed to develop and implement a comprehensive GIS cannot be underestimated. The task of geocoding all the Council's social, demographic and economic information together with the geocoding of land parcels, road networks, water supply and sewerage networks will take a considerable amount of time and resources. Nevertheless, we would emphasise that GIS development in the Regional Council is being driven by the strategic and managerial needs of the Council. During the year 1989-90, practical GIS applications were being implemented in five departments within the Regional Council in order to evaluate the potential of GIS in the context of services using different types of information and with differing management information needs. These developments were, however, taking place within a corporate approach designed to ensure that the corporate information systems design objectives referred to above were being achieved.

GIS development in Birmingham City Council: review of a pilot study

The Birmingham GIS Pilot Study began in early 1988 and had two objectives: the first was to assess the potential value of the development of GIS and digital mapping applications to the Authority as a whole and to individual departments in meeting both policy and operational objectives, and second, to produce a framework for the future handling of GIS within the City Council. The pilot exercise progressed along two separate channels: the objective of the first channel was to identify and test a series of practical applications and the objective of the second channel was to examine systematically the potential for the development of GIS throughout the City Council (this second channel is essentially concerned with educating potential GIS users about the benefits of GIS). The benefits of GIS to local government are potentially considerable as it has been shown that about 80 per cent of information in local government is spatially related (Bromley and Coulson, 1989): the objective in Birmingham was to quantify some of these potential benefits.

In order to get maximum corporate benefit from the first channel of the pilot study, test application areas were selected according to how well they satisfied several criteria. The first selection criteria was that the test applications should incorporate point data, zonal (polygon-based) data and network data. Second, the cost of data capture in GIS development is substantial (a recent study has shown that 60-75 per cent of the total development costs of GIS are consumed by data base creation — see Maffini, 1990) and in order to gain an insight into the relative costs of data capture, test applications were selected which would use a variety of data collection methods (these included mainframe to PC and PC to PC data transfer, manual transfer and digitising). Third, while mapping is important and often useful in improving our understanding of relationships, it is often not essential — indeed, many of the potential users of GIS are not 'map literate' — and so, test applications were selected where an advanced mapping facility was desirable but not strictly necessary. It is also important to stress that the system is location-orientated primarily because this is often the most common and stable item of information associated with almost all local government activities, but the geographic tag does not imply that the spatial attribute is the most important information item. Fourth, because different types of users have different needs from a GIS, it was decided to select test applications which reflected differing end-user requirements. Three generic application types were identified: these were policy formulation (i.e. strategic level use); operational/resource allocation (i.e. geared more to the needs of facility/functional managers) and corporate data sharing operations (i.e. geared to developing corporate information synergy). Finally, test applications were selected which would allow us to test whether 'generic' application code could be written which was transferable within and between application areas. Given the size and multi-functional nature of local government it was thought that substantial savings could be made, and subsequent applications could be developed more quickly, if a library of user-friendly generic routines could be established. Also given the size of the City of Birmingham, it was decided to limit the pilot study to one area of the city: the area chosen was the East Birmingham Urban Development Area which is politically a high profile area subject to rapid development.

Eight test application areas were eventually selected and information about each of them is presented in Table 6.3. The test applications, because they were selected to represent a wide range of applications types in local government, provided some valuable information to develop a framework for the more widespread application of GIS. The decision to develop a library of generic applications meant that initial progress was slow but that once basic routines had been developed the pace of progress accelerated. The overriding conclusion was, that by taking a strategic approach, GIS is capable of wide-ranging application throughout local government at the corporate level, the interdepartmental level and at the departmental level. However, it must be emphasised that GIS is not about clever mapping or clever computing — it is all about adding value to existing information to improve the quality of decision-making. GIS are primarily concerned with information and not maps: maps should be viewed simply as another source of information and as a means of aiding the presentation of managerially or strategically relevant information.

Table 6.3   Birmingham GIS pilot study: application attributes

Application	Point	Spatial aspects Polygon	Network	Data capture Mainframe	Micro	Manual	Maps	Text
Census		x		x				x
Land resources	x	x			x	x		x
City Terrier		x				x	x	x
Bridges data base	x		x		x	x	x	x
Lighting data base	x				x	x	x	x
Sewers data base			x		x	x	x	x
Educational transfers	x	x				x		x
Library stock monitoring	x	x			x			x

Source: Gault and Davis, 1988

Some practical lessons and guidelines for the adoption of GIS in local authorities

 Introducing GIS to local government

Throughout this chapter, we have suggested that GIS is potentially a powerful tool with which to manage the increasing amounts of information which are now required to manage the changes currently

facing local government: if this is the case, why have they not been more widely adopted in UK local government? The concept of GIS is hardly new or revolutionary in UK local government (see Willis, 1972) and developments in this area can be traced back as far as the GISP Report of 1971 (*General Information System for Planning*, DOE, 1971). Ironically, it may be that these past experiences have prevented the more rapid adoption of GIS, as they have led to problems of association and definition. To many potential users in local government, GIS are too simplistically associated with map handling, routine 'technical' activities (e.g. engineering, land-use planning) and routine land and property management issues. They are not generally seen as broader decision-support tools which provide a medium for the more efficient integration of disparate data sets or as integral components of strategic planning and management procedures.

These associations immediately limit peoples' perceptions as to whether GIS is relevant to their particular interests and, as a result, the subject tends to be approached on the basis of answering the question 'what can I use GIS for?' rather than asking the more profound question 'what sort of information systems and analytical procedures do I need to enable me to manage my business more efficiently?' If this latter approach is adopted, the focus for GIS development then becomes one of improving the process of managing change through the improved use of available information assets. In order to increase the use of these information assets it is essential to be able to integrate and manipulate information with the maximum feasible flexibility. Changing the way that GIS are perceived, particularly at the senior management level of an authority, is essential if GIS is to play a stronger role in the management of local authorities and in the broader management of change.

Implementing GIS in local government

If we assume that this initial educational hurdle can be overcome, there are still several issues which must be addressed if GIS is to be successfully implemented in local government. These issues can be grouped under the four headings: developing a GIS strategy; information management; applications justification; and organisation and management.

DEVELOPING A GIS STRATEGY

It is our view that most 'GIS Strategies' in local government tend to be either technology-led or too narrowly centred on a specific process or task: they are often based on the question 'how can I use GIS to "computerise" what I do now'? Given our view that the overriding requirement is to improve the ability to manage change, then we would suggest that concentrating on how activities are currently carried out, or the degree of technological functionality currently required, is not the most important strategic consideration. Rather, we argue that any GIS strategy must have the following qualities. First, the strategy must be business-led and not technology-led and the focus must be on addressing the needs of the business as a whole, not on what current (and anticipated) technology (which is subject to rapid change) will allow. Second, it must be policy-led and not process-led, as a concern with why the authority is doing something and its impact is a more important issue than examining the minutiae of current administrative processes. Third, it is essential to adopt a corporate approach if a degree of synergy is to be realised which will assist both the management of change and increasing operational efficiency. Fourth, the strategy must be actively endorsed by senior management and viewed by them as an integral part of the management process. This is also important from the point of ensuring the adoption of a corporate approach.

A central implication of the above is the need for authorities to distinguish between an IS (information systems) strategy and an IT (information technology) strategy. An IS strategy addresses the issue of what information systems are needed to support the pursuit of the business objectives, whereas the IT strategy addresses the question of the technological framework needed to deliver the IS strategy.

INFORMATION MANAGEMENT

It is suggested that the sophisticated management of information itself is required if a GIS approach is adopted. However, the issue must be addressed in terms of what is feasible given current skill, organisational and technical constraints. It is essential that information is seen as an asset within the organisation in its own right, but it is equally important that the processing of information, including its collection and storage, is viewed as a cost. Therefore, there must be explicit consideration given to the balance of costs and benefits between the needs of the users and providers of information. In the case of GIS specifically, the relative roles of maps and text need to be made explicit.

Although it was stated that a corporate approach was essential, this should not be confused with central control. Central control of information processing is perceived as being unresponsive, unwieldy and expensive. A distinction has to be made between the business processes which generate the information, and the uses of that information by the organisation at large. It is essential to allow departments to control their own administrative processes within a framework coordinated centrally.

To many people, standards in information management cost money and are simply unnecessary constraints and irritants. This view is often correct. Consequently, standards should be kept to a minimum if widespread participation is desired. For GIS, it is critical that a convention on locational referencing is adopted and that rules on the use and updating of the map base are accepted. There will also be a need to establish a basic data definition dictionary. Even these minimal standards will have to be presented in such a manner as to make it clear that all participants will benefit by adopting them.

The accuracy and currency of data is vital as it will determine system credibility and ultimately its degree of use and usefulness. However, accuracy is not an absolute quality and it means different things to different people in different circumstances. The first step in addressing this issue is to ensure that the 'ownership' of, and responsibility for, items of information are clearly specified and understood. Once this has been established, negotiations can take place between the users of the information and the owners of the information as to the level of access, degree of accuracy or tolerance and the currency of the information. These negotiations should form the basis of a clearly understood 'contract' between the provider and users. A distinct problem in this area is the generally poor quality or inappropriateness of nationally available statistical material for local authority policy-making (Worrall, 1988), though, as in the Strathclyde example, it has been shown that these problems can be overcome given the necessary vision and commitment.

Although it has been argued that an authority's strategy should not be technology-led, technology is obviously a factor which has to be considered. This is not an appropriate vehicle for a detailed discussion of technical issues, although certain broad criteria can be laid down: flexibility and integration between IT systems is essential; information processing constitutes a business cost and solutions should seek to minimise costs to users and providers; and it is imperative to concentrate on the data base aspects rather than graphics as graphics facilities can be added later while redesigning large data bases is expensive. The communications aspects of IT will become more important and this must be seen as a central element of any strategy rather than an afterthought.

APPLICATIONS JUSTIFICATION

While we are of the view that the introduction of GIS into a local government environment is desirable for strategic reasons, there are many examples which show that local government does not have a good track record of investing in strategic information systems and has tended to concentrate on discrete process-orientated applications. One of the advantages of a GIS is that its development can yield significant short-term operational benefits at the process level while the broader corporate uses are still within their gestation periods. It is possible to initiate the implementation of GIS by identifying discrete applications which will have substantial benefits. However, it is important to maintain a balance between strategic and administrative or procedural uses and not allow administrative uses to dominate the more critical longer-term strategic considerations.

Local government is not short of potential application fields for GIS but the basic problem of unwillingness to invest in IT for strategic management will remain a problem until the key issues and relationships we have indentified are addressed by senior managers and elected members in local government. It is essential that senior managers and strategic level decision-makers become aware of the real costs and lost opportunities of the 'do nothing' option and begin to see the development of strategic information systems as an investment leading to more effective decision-making. For these issues to be addressed will require fundamental changes in the methods and accounting frameworks currently used to assess the costs and benefits of IT projects and this in turn will require changes in the prevailing management style and approaches used in UK local government.

ORGANISATION AND MANAGEMENT

The issues in this area need to be addressed from two viewpoints: first, in terms of identifying the necessary conditions for the successful introduction of GIS as a strategic management tool, and second, in terms of the longer-term impact on the organisation and its management style of adopting a GIS-centred approach to information systems use. The creation of GIS can be expensive in both time and money, particularly in the early stages of development. If authorities are to invest in them, they must have a management culture which is pro-active, change-orientated and in which the management of change is seen as the dominant managerial consideration. They must also have a culture which is conducive to the effective use of, and investment in, information: this culture will be one in which information is regarded as an asset in its own right, rather than simply a by-product of routine administrative processes.

In the longer term, the impact of GIS development on local authority organisations could be fundamental. The improvement in communications, which would be a necessary precursor of wholesale information integration, would offer opportunities for the creation of a more physically and functionally decentralised organisation: this is indeed already a trend within local government. The increased access to a variety of departmental systems from a single access point could greatly enhance (controlled) public access to relevant systems and be the cause of a shift in the prevailing local government culture from administration towards greater public involvement and the greater democratisation of decision-making (Hancox *et al.*, 1988).

Conclusions

In this paper, we have taken a slightly different perspective for examining the potential for GIS development in local government. Rather than focusing on the more technological and technocratic aspects of GIS, we have tried to examine the political, organisational, managerial and policy-analytical environment in which GIS will have to exist. It is our firm view that it is essential to understand the processes and forces which are affecting organisations as a whole before any sensible recommendations can be made about the application of GIS in UK local government. In our two case studies we have shown that local government is a fertile environment for the implementation of GIS but that there are real problems and impediments which must be overcome if it is to achieve its full potential — most important, the problems are as much organisational as technical. Overcoming the lack of awareness of the potential of GIS at the senior level is perhaps the greatest task that needs to be achieved: like Scholten and van der Vlugt (elsewhere in this volume), we believe that the interaction between senior management in local authorities, information specialists and policy analysts will prove to be critical.

It is clear that local authorities in the UK are in the middle of the most concentrated period of change experienced this century. The changes are not restricted to the impact of legislation but extend to address the very role of local government in society. As a result, local authorities will have to develop a management culture which is change-orientated and sees the management of change as being of central concern. We have argued that this shift can only be achieved if local authorities adopt more rational planning procedures and that these procedures depend on local authorities developing an information culture where information is viewed as a valuable resource in its own right. Such a culture would place emphasis on the ability to integrate information on an extremely flexible basis in order to realise synergistic benefits.

Because of the nature of local government services and products, and because programme implementation and service delivery tends to be area based, we believe that local authorities should adopt an IS strategy based on integration through the use of spatial referencing, and that future IT strategies should be designed around that single principle. However, we have argued that the development of GIS will only be successful if it is business-led and takes account of the realities of organisational politics. Finally, it is our view that in order to develop the capability to manage change, the senior management of local authorities must adopt a much more active role in the development of IS and IT strategies, though a substantial educational programme will first be needed to make them aware of the contribution that GIS can make to the more effective and efficient management of change.

Bibliography

Beaumont, J. (1989), 'Market analysis, commentary', *Environment and Planning A*, 21, pp.567-70.

Black, R. (1985), 'Instead of the 1986 Census: the potential contribution of enhanced electoral registers', *Journal of the Royal Statistical Society*, series 'A', 148(4), pp.287-316.

Bromley, R. and Coulson, M. (1989), 'The value of corporate GIS to local authorities: evidence of a needs study in Swansea City Council', *Mapping Awareness*, 3(5), pp.32-5.

Clapham, D. (1989), 'The new housing legislation: what impact will it have?', *Local Government Policy Making*, 15(4), pp.3-10.

Derbyshire, M.E. (1983), 'The application of statistical methods in the personal social services: a review', *Journal of the Royal Statistical Society*, series 'A', 146(2), pp.115-49.

DOE (1971), *General Information System for planning*, Report of the Joint Local Authority, Scottish Development Department and Department of the Environment Study Team, Department of the Environment, London.

Flynn, N. (1989), 'The "new right" and social policy', *Policy and Politics*, 17(2), pp.97-109.

Gault, I. and Davis, S. (1988), 'The potential for GIS in a large urban authority: the Birmingham City Council Corporate GIS pilot', *Mapping Awareness*, 2(5), pp.38-41.

Griffiths, R. (1988), *Community care,* Department of Health and Social Security, London.

Hancox, A.P. (1988), 'Developing corporate strategy: a framework for local government', *Local Government Policy Making,* 15(1), pp.60-4.

Hancox A.P., Worrall, L. and Pay, J. (1988), 'Developing the customer orientation: the Wrekin experience', *Local Government Studies,* 15(1), pp.16-25.

HMSO (1972), *The new local authorities: management and structure* (the 'Bains' Report), HMSO, London.

HMSO (1986), *Report of the committee of inquiry into the conduct of local authority business* (the 'Widdicombe' Report), Cmnd 9797, HMSO, London.

John, P. (1989), *Introduction of the community charge in Scotland,* Local and Central Government Relations Research Programme Report no. 1, Policy Studies Institute, London.

Maffini, G. (1990), 'The role of public domain databases in the growth and development of GIS', *Mapping Awareness,* 4(1), pp.49-54.

Pay, J. (1989), 'An action-learning framework for the targeting of resources into social priority areas', *Research and Intelligence News,* no. 32, pp.2-3.

Peutherer, D. (1988), 'Developing small area information systems for service and resource planning in Strathclyde Regional Council', Paper presented to the Fifth Regional Science Association International Workshop on Strategic Planning, ITC, Enschede, The Netherlands.

Smith, W. (1989), 'Information society: fact or fiction', in P. Shand and R. Moore (eds), *The Association for Geographic Information Yearbook — 1989,* Taylor and Francis, London.

Strathclyde Regional Council (1980), *A social strategy for the eighties,* Strathclyde Regional Council, Glasgow.

Strathclyde Regional Council (1988), *Strathclyde — generating change,* Strathclyde Regional Council, Glasgow.

Willis, J. (1972), *Design issues for urban and regional planning information systems,* Centre for Environmental Studies WP71, Centre for Environmental Studies, London.

Worrall, L. (1988), 'Information systems for urban and regional planning in the UK: a review', Paper presented to the Fifth Regional Science Association International Workshop on Strategic Planning, ITC, Enschede, The Netherlands, *Environment and Planning 'B'* (forthcoming).

Worrall, L. (1989), 'Urban planning processes and strategic information systems', Paper to the International Conference on Computers in Urban Planning and Urban Management, University of Hong Kong.

Young, R. (1989), 'Lessons from Strathclyde's experience: boosting peoples' self-confidence', in K. Dyson (ed), *Combating long-term unemployment: local/European community relations,* Routledge, London.

Chapter Seven

Elements of a model-based Geographic Information System for the evaluation of urban policy

Mark Birkin, Graham Clarke, Martin Clarke and Alan Wilson

Introduction

In this chapter, our objective is to reflect on the progress of a research programme at the School of Geography at the University of Leeds which has a history of nearly two decades of involvement with urban and regional analysis. This interest was initially focused around the application of model-based methods, and this approach always had a strong computational component. This interest in computer-oriented methods has led the research naturally, along with the research of many others, into a concern with the use of Geographic Information Systems in the analysis.

We believe, however, that our more traditional analytical modelling interests have served to give our applications of GIS a distinctive flavour, with at least two kinds of unusual characteristics. The first of these is that our approach to GIS remains model-based, rather than data-based, as are many such systems. The elements of the modelling approach are reviewed in the second section of this chapter. Secondly, we are concerned with the development of customised information systems using a variety of software products. This strategy yields greater flexibility and power than the use of single, 'off-the-shelf' GIS packages such as ARC/INFO. Further discussion of this methodology is also provided in the third section of this chapter.

Much of our theoretical contribution has been elaborated with respect to the Leeds and West Yorkshire urban and regional economies. Another feature is that much of this work has involved close collaboration with various local bodies which require an analytically-based foundation for policy development. These bodies include the Department of Industry and Estates of Leeds City Council, the Leeds Chamber of Commerce and the Yorkshire and Humberside Development Association. One of the changing emphases of recent years throughout the UK has been away from traditional town planning towards a greater concern with urban economic development. This has been reflected in the application areas we have chosen, and in a later section of this chapter we describe the results of our work in relation to labour markets, retailing, housing and education.

Model development

Modelling strategy

At the heart of our activities has been the development of two kinds of comprehensive model of the urban and regional economy. The first, concerned with microsimulation, is the most detailed, but demands too much data and effort for some applications. The second represents state-of-the-art spatial-interaction based modelling. The models we have built are outlined below.

The microsimulation model

We shall see in the next subsection and in the section on applications that aggregate model-based analysis is a valuable tool for many kinds of spatial development and evaluation issues in the urban and regional context. However, there is often a demand for information at much finer levels of resolution, in particular to look at the distributional effect or impact of policy as it relates to different types of individuals or households. In order to be able to serve these kinds of needs, there is a need for micro-data on the attributes of individual persons and of household units.

One of the problems here is that the kind of micro-data which we need is basically unobtainable in the UK for either demographic, social or economic characteristics though this is not the case in some countries (for example, in Scandinavia due to the existence of extensive register-based information systems — see Thygesen, 1984). However, a solution to this problem can be found by generating this information synthetically from aggregate distributions, that is to produce imaginary but 'typical' sets of individuals generating a population which is consistent with known aggregate distributions. This approach is technically and computationally demanding, but presents the analyst with the basis for information systems of unprecedented power, because it provides us with a framework for linking together a wide variety of data bases that are usually only loosely connected, and often not explicitly related to small geographic areas. The major inputs to the synthetic data generation procedure in our studies are the 1981 Census of Population and Households and national government surveys (such as the Family Expenditure Survey, the General Household Survey and the New Earnings Survey) though *ad hoc* local surveys on shopping behaviour and journeys to work have also been used and form an integral part of our data system.

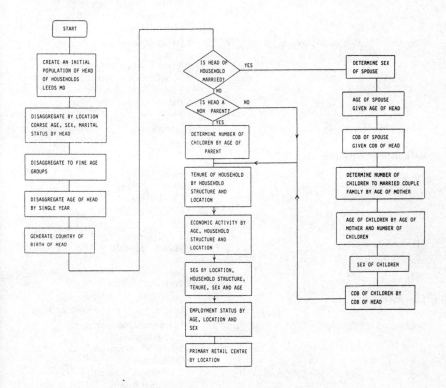

Figure 7.1 *The sequence of steps in attribute generation*

Table 7.1 Attributes generated in the microsimulation model

Attribute	Number	Details
a) Household attributes location		
Location (Census Enumeration Districts)	1565	1 DAAA01 2 DAAA02 ... 1565 DABK47
Household structure and composition	5	1 Single person, retired 2 Single person, not retired 3 Married couple, no children 4 Lone parent family 5 Married couple with children
Tenure	3	1 Owner-occupied 3 Other 2 Council rented
Country of birth of household head	7	1 Great Britain 2 Eire 3 New Commonwealth — India 4 New Commonwealth — Caribbean 5 Rest of New Commonwealth 6 Pakistan 7 Rest of world
Primary retail location	52	1 Hemsworth 2 Normanton ... 52 Bradford
b) Individual attributes		
Status within household	5	1 Head 2 Spouse of head 3 Child of head 4 Other dependent 5 Other
Exact age	86	0 1 2 ... 85+
Sex	2	1 Male 2 Female
Marital status	3	1 Single 2 Married 3 Widowed/divorced
Economic activity	4	1 Inactive 3 Retired 2 In work 4 Seeking work
Socio-economic group	7	1 Employers and managers 2 Professional 3 Intermediate non-manual 4 Skilled manual 5 Semi-skilled manual 6 Unskilled manual 7 Other
Industry	7	1 Agriculture 5 Distribution 2 Energy 6 Transport 3 Manufacturing 7 Other 4 Construction
Exact income	1000	0 1 2 ... 999+
Household expenditure by category	8	1 Housing 5 Household 2 Fuel 6 Other goods 3 Food 7 Transport 4 Clothing 8 Services

An overview of the procedure adopted is provided in Figure 7.1. Data is generated on a step-by-step basis by sampling from aggregate distributions, initially on the demographic composition of the population, and then moving on to various socio-economic attributes and activity patterns. At the end of the chain, income and expenditure are introduced into the procedure.

A full set of attributes generated in this way is shown in Table 7.1. The usefulness of this data can be demonstrated in a number of ways. First, we can reaggregate micro-level data to estimate aggregate distributions which were not previously known. For example, it is possible to examine a range of household characteristics related to the socio-economic characteristics of their members. Second, it is possible to generate small area information from spatially aggregate data — this is particularly valuable for the generation of small area income and expenditure profiles. Third, we can combine micro-data with macro-level models, such as a model of retail trip-making. This allows us to produce detailed catchment area profiles for shopping centres (or labour markets). Finally, we can use a variety of procedures to 'update' micro-data to make it more representative of changes since the various data were collected, and to make forecasts about future trends. A variety of applications of this model are described later in the chapter.

The design of an SI-based comprehensive model

Our comprehensive urban model is based on the interdependence between three different subsystems of the urban economy — population and housing, services and the economic base. In relation to each of the three subsystems, there are in turn three separate kinds of focus. The first is an interest in structure — here we are concerned with the nature and characteristics of the housing stock, of employment opportunities, or of patterns of service provision within small areas. The second focus is an interest in activity patterns, which typically means that we want to concentrate on how particular groups of people (such as the residents of a small area in the city) use services and what kinds of job they have. In practice, this means that we are concerned with processes of spatial interaction within a region. To a large degree, it is these processes which generate the need for a comprehensive approach, because they involve interactions between the subsystems. We are also concerned with other kinds of spatial interaction processes which do not link subsystems, specifically migration and inter-industry trade flows.

The third focus is a concern with dynamics — that is the way that spatial structures or activity patterns can be expected to develop in time. Some of the dynamics will be generated endogenously within the system: hence we can expect natural growth in the population, coupled with economic responses to supply-side imbalances in the manner of Harris and Wilson (1978). This approach to stock dynamics is equally applicable to the economic base or housing construction as it is to the retail sector (see Birkin, 1990, for example). A second kind of dynamic process is the response to exogenous change: for example, how does the urban system readjust after a major firm relocates to the city with the generation of 2,000 new jobs? Given our concern with estimating the stocks of various phenomena within small areas and our concern with estimating and modelling spatial interactions, the application of GIS techniques is clearly an essential element to the further development of our research.

The main structure variables of the model can be described as follows:

H_i^{rs} — housing in zone i, of tenure r and 'type' (size, quality etc) s;

P_i^{abcd} — population in zone i, of age a, sex b, marital status c, and social class d;

W_j^k — level of provision of a service type k in zone j;

E_j^{dg} — employment in zone j of persons of occupation d, in industry g;

Then we have the following sets of spatial interactions:

1. the use of services by the population;

2. journey to work, and the demand for housing;

3. population migration between small areas;

4. inter-industry trade flows.

The endogenous dynamics combine both the structural and activity components:

1. population change — this is a combination of natural increase and migration;

2. housing change — this is generated by supply/demand imbalances in the housing market, mediated through the price mechanism. Thus, where there is pressure on housing stock, there will be price inflation but also a tendency for the stock to expand, and vice versa;

3. changing service provision — this comes about as a response to profitability in provision: where revenues exceed costs then service levels will tend to rise; where costs exceed revenues then levels will contract;

4. employment growth — this can also be related to the economic
    success of the activity in question, although in practice employment
    change is more likely to be exogenously determined either by
    expansion, contraction or relocation. *Ad hoc* processes will also
    impinge on the service and housing sectors as developers implement
    particular decisions for new house builds or shopping centre
    development. These types of decision are susceptible to impact
    analysis, as we shall see below, while the more general patterns of
    growth or contraction in response to supply-demand or revenue-cost
    imbalances can be used more generally to project changing system
    structure.

A fourth level of calculations involves the computation of performance
indicators. These indicators can be used to summarise the features of
complex spatial interaction processes, or may be combined for the
evaluation of model outcomes, as we aim to show in the section on
applications below.

The comprehensive model operates in three ways: these are the static,
comparative static and dynamic modes. In the static mode of analysis we
are concerned with estimating missing information, such as spatial
interaction patterns, and with the calculation of performance indicators.
In comparative static mode, we can assess the impact of changes to the
system — such as the opening of a new shopping centre or the relocation
of a manufacturing plant. Finally, we can use dynamics to forecast
developments in spatial structure or activity patterns — for example to
answer questions such as where will the demands for new housing be in
ten years' time?

As we have seen above, it is possible to use model-based methods to
link together a variety of data sources. However, the comprehensive
model can only be constructed and calibrated using data which is
generally available and easy to obtain. Of course detailed information on
population, housing and economic activity is all available from the census
(although the business of updating these kinds of data base is a more
complex one, and again it may be appropriate to address this problem
using microsimulation methods). Patterns of industrial, retail and service
supply can all be established using a variety of directory sources (such as
Newman's Retail Directories or Dun and Bradstreet's Industrial
Directories). The demand for retail and service commodities can be
estimated by combining demographic data with Family Expenditure
Survey or General Household Survey estimates of household
expenditure patterns.

Information Systems as Integrators

Information strategy

We have explained in earlier sections why we have chosen to give a higher profile to our information systems research. Here we need to explain our strategy in more detail. Consider a number of propositions. We have already indicated that we need to assemble large amounts of data from diverse sources as a basis for model-building. Access to this data is often useful for analysts or planners. Model outputs are another kind of 'data' or — from now on — 'information'. This is particularly true of the microsimulation model since part of its function is to generate information which fills in 'gaps' in data — such as small area income distributions. We can combine elements of data and information in various ways to construct a great variety of performance indicators. Again, analysts and planners need access to these. It is often not possible to pre-specify what the detailed information needs of a particular planning task are — and new needs are usually identified as the process proceeds. An essential aim of model-based planning is to be able to run models to generate performance indicators to test the impacts of plans. Planners and analysts often need to do various kinds of statistical analysis — on either data or model-generated information — and so analysts need good graphical presentation of the results of such analyses and so need good mapping facilities.

The tasks implied by any one of these propositions as part of a particular planning undertaking are nearly always carried out separately. We have made it a key feature of our research to use the idea of an information system to integrate all these tasks into one comprehensive system. This takes us a long way beyond the traditional concept of a GIS — if only because we want to have on-line access to model-based analysis. We have been able to develop the software to achieve such an integrated system, or what we term an 'IGIS' — that is an Intelligent Geographic Information System. The structure of the IGIS we have developed is shown in Figure 7.2.

The development of this system has been facilitated by the increasing power of PCs (which has removed the need for a mainframe for much of the modelling) together with the availability of software which makes it possible to combine subsystems written in different languages. In the system we have developed in our research programme, we have used DBASE III for the information system, user controller and report generator, graphics and statistical analysis; FORTRAN for the

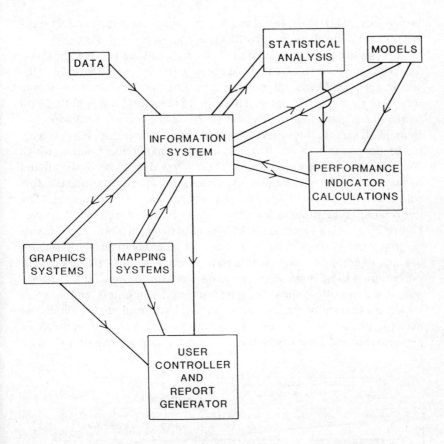

Figure 7.2 *The structure of an intelligent geographical information system*

modelling; and QUICKC for the mapping system. It is the overall system which can be described as an IGIS.

The West Yorkshire IGIS

In the present implementation, the system contains information on five different kinds of urban subsystem: these are population and housing, the provision of urban services, the utilisation of services, the local economy and the local labour market. A range of performance indicators can be analysed for each subsystem. Altogether, 44 different performance indicators are considered, but all of these may be disaggregated further by age, sex, social class, industry of employment, or retail product types as appropriate to the particular needs of the user.

This 'sectoral' focus is combined with spatial analysis of the West Yorkshire region which can be disaggregated by electoral wards (the smallest politically accountable unit in the UK), postal districts, postcode sectors or metropolitan areas. There are 127 electoral wards in the West Yorkshire region, and nearly 300 postal sectors. The combination of spatial and sectoral interests starts to generate a large data base. Access to the data base is controlled by a series of hierarchical menus which comprise the information system. These allow data to be retrieved and displayed: for example, Figure 7.3a shows a list of performance indicators which can be selected for the population and housing sector. The information system also allows the data to be structured: for example, Figure 7.3b shows a more complex situation in which wards in the Leeds Metropolitan District have been ranked by the mean journey-to-work distance of employers and managers (in Column 1). (Note that for ease of presentation these illustrations are shown as monochrome tables of data, but would usually appear as multi-coloured computer screens.) An important feature of the data base retrieval system is that it allows the user to generate his own tables as combinations of a large number of performance indicators, which is extremely useful for exploratory data

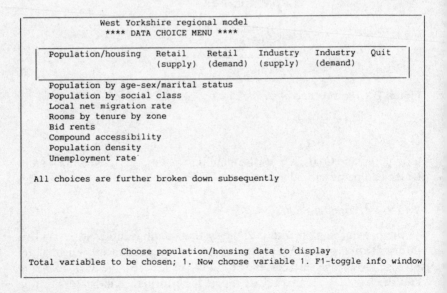

Figure 7.3a *An example of the West Yorkshire IGIS: data choice menu*

analysis. In systems where the data or information on offer is so rich, it is difficult to offer useful analytical procedures which are more than exploratory, although bivariate statistical analyses such as correlation and regression are available and more complex multivariate methods could be added relatively easily if these were seen as necessary.

```
                West Yorkshire regional model information system
 Area is:LEEDS                                   Model is Baseline (1981)
Ward                 TRIP       TRIP      TRIP      TRIP   Hans.acc  Eff.del
                Emps/mang  Non-manl    Sk.man    SS.man  Emps/mang   SS.man
                 All.ind    All.ind   All.ind   All.ind   All.ind  All.ind

      Wortley       6.6        4.7       3.7       3.9     4591.9   2942.6
      Wetherby      4.0        3.3       3.1       3.0     8725.0   7362.6
      Bramley       3.5        3.4       3.0       3.2     9027.7   8062.6
      North         3.4        2.9       2.6       2.8    12591.3  11613.5
      Cookridge     3.2        3.1       2.9       2.9    13658.6  14891.8
      Hunslet       3.2        3.1       3.0       3.0    14752.9  14075.8
      Garforth&Swil 3.0        2.9       2.5       2.7    15119.6  15827.0
      Middleton     3.0        3.0       2.8       2.8     9682.3   8845.2
      Beeston       2.9        2.7       2.8       2.8    12699.1  10872.5
      Moortown      2.9        2.6       2.4       2.5    16115.2  16368.8
      Roundhay      2.9        3.2       3.2       3.1    11307.5  10347.8
      Horsforth     2.8        2.6       2.5       2.5    16002.3  17627.2
      Harehills     2.7        2.6       2.6       2.7    16359.2  14753.0
      Burmantofts   2.6        2.4       2.3       2.3    12534.7  15539.0
      Whinmoor      2.4        2.3       2.1       2.0    11099.5   9973.7
      Morley-North  2.4        2.2       2.0       2.0    16771.3  13446.5
Item:1 1-34 of 34                        Ward summary:highlight ward, PRESS ENTER
_F1 - full variables, F2 - hotkeys_____Pg Up,Pg Dn to scroll, Esc to Escape
```

Figure 7.3b *An example of the West Yorkshire IGIS: some key indicators*

The data base outputs can then be analysed geographically. Figures 7.4-7.9 inclusive are all examples of the kinds of maps that can be produced (again, it must be borne in mind that these will usually appear as multi-coloured computer output and that the technology now exists to produce hard-copy colour maps at relatively low cost). These maps can be rescaled or annotated, but it is important to note that the mapping element is not necessarily the heart of our IGIS, as it is with many Geographic Information Systems. Rather, mapping is seen as only one of a range of core activities, of which others are the flexible combination of performance indicators, the identification of exceptional geographies (through ranking or clustering of the data) and the identification of areas with common features (through filtering — for example, filter all areas with an unemployment rate between 6 per cent and 7 per cent).

So far, all the activities we have considered are baseline functions, which relate to the census year 1981. The modelling capability of the system allows us to add a temporal dimension to the analysis. Thus we can 'project' what is happening to the activities in the system in 1986, 1991 and so on at five-year intervals. Alternatively, we can change a limited number of activity levels in the system, recalculate the model equations, and look at a new set of performance indicators — a 'comparative static' analysis. For example, we could add a new shopping centre and see what effect this has on local accessibility to shops (see the section on applications). Or we can combine the comparative static and dynamic options to see what happens if we 'fix' some sectors and forecast the others with a free response. For example we can experiment with questions such as: What happens if we build 500 new houses on a development site? Where will the residents find jobs in the short, medium and long term, and what impact will this have on the utilisation of local services? What happens if we build 1,000 houses in a particular area, or what would happen if we developed the land as commercial offices instead?

Applications

Local labour markets

It is clear that the last decade has witnessed a dramatic restructuring of many regional and local labour markets, with service sector activities now being more dominant than traditional manufacturing sectors. One of the consequences of this restructuring has been a large increase in the numbers of unemployed, particularly in certain areas of the city — usually the inner city. The aims of our research in local labour market analysis have been twofold. The first aim has been to draw upon the microsimulation Leeds data base (in conjunction with models calibrated for the journey-to-work patterns in the city) in order to break down the aggregate picture of (un)employment. This has enabled us to incorporate attributes such as socio-economic group, age, sex, race and industry type with household structure and composition variables. The second aim has then been to focus on the journey-to-work models and demonstrate the disaggregate impact of different types of job creation programmes. In particular, we have been interested in how new jobs have (and will) affect unemployment levels in certain areas rather than merely the balance of migration and commuting flows.

The detailed picture built up on the demand side of the labour market has provided a powerful framework for analysis and has put us in a much more effective position to try and understand the functional relationships within the Leeds local labour market. It is useful to conceptualise the

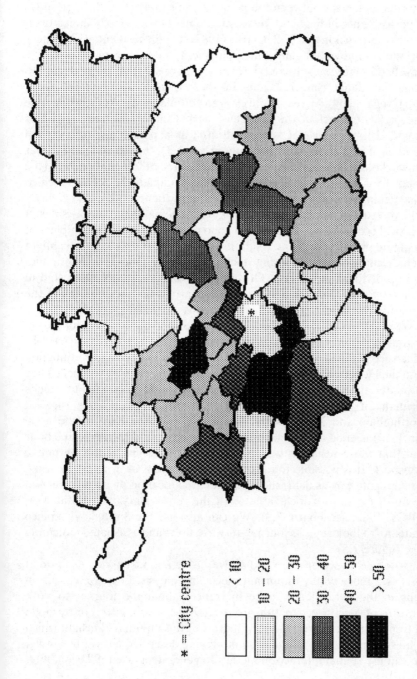

* = City centre

< 10

10 — 20

20 — 30

30 — 40

40 — 50

> 50

Figure 7.4 *Leeds clerical service workers commuting by bus*

interdependencies between the demand and supply sides through suites of performance indicators. The examination of these sorts of indicators is described in Birkin and G.P. Clarke (1987). On the residential side, these include employment and unemployment rates, average distances travelled, degree of self-containment within areas (that is the degree to which residents can find jobs in their own locality) all of these disaggregated by age, sex, social class and ethnic status. At the workplace end, indicators include the 'degree of market share' (the number of jobs provided in a particular workplace area as a percentage of the total number of jobs across the whole labour market), the size of market areas for workplace zones and, once again, the degree of self-containment (the opposite of the previous residential indicator, namely the degree to which a particular workplace zone employs from residents within that zone).

Given such a detailed picture of both the demand and supply sides of the local economy, we are now in a good position to model the impacts of changing either side of the relationship. On the supply side, this might include changes in the housing stock (by locality), in local demographics or in individual skill levels. On the demand side, we are interested in changes to the number and type of job provision and in defining the potential catchment areas for any new jobs created.

We have recently been able to use the modelling and information system to monitor the impacts of a possible 2,000 new Civil Service jobs which are planned to be relocated from London to Leeds, in conjunction with the Department of Industry and Estates at Leeds City Council. They estimated that around 500 new clerical jobs would be created for Leeds residents if the city was successful in attracting the jobs (against Nottingham and Manchester). The Department of Health (the body involved) needed to be persuaded that the city could recruit such labour and that those residents were well served to the city centre by public transport. It is possible to make a certain amount of progress by using disaggregate census data, such as residential location by occupation and mode of travel to work (Table 47 of the census Small Area Statistics; OPCS, 1982; see Figure 7.4). We can also use the Special Workplace Statistics to look at gross journey-to-work flows and associated indicators (see Figure 7.5).

However, as we have argued above, using a model allows us to do at least two more vitally important pieces of analysis. First, we can fill in gaps in the data, hence make inferences about the journey-to-work patterns of specific occupation groups (by age, sex, industry, or by mode of transport if necessary) — see Figure 7.6. It is this kind of insight which comes from the microsimulation modelling described above. Secondly, we can use spatial interaction models to generate a 'what if?' capability,

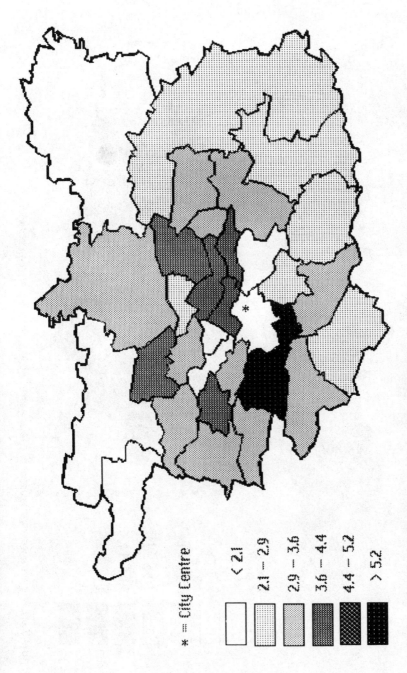

Figure 7.5 *Percentage of city-centre workers resident in Leeds wards*

* = City Centre

< 2.1

2.1 — 2.9

2.9 — 3.6

3.6 — 4.4

4.4 — 5.2

> 5.2

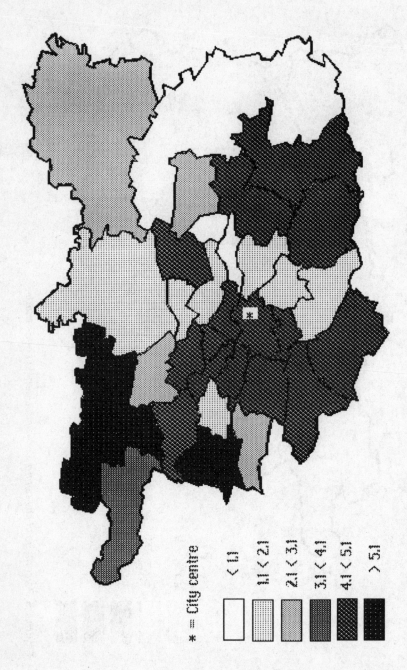

Figure 7.6 *City-centre workers resident in Leeds wards: services — clerical (percentage)*

* = City centre

< 1.1

1.1 < 2.1

2.1 < 3.1

3.1 < 4.1

4.1 < 5.1

> 5.1

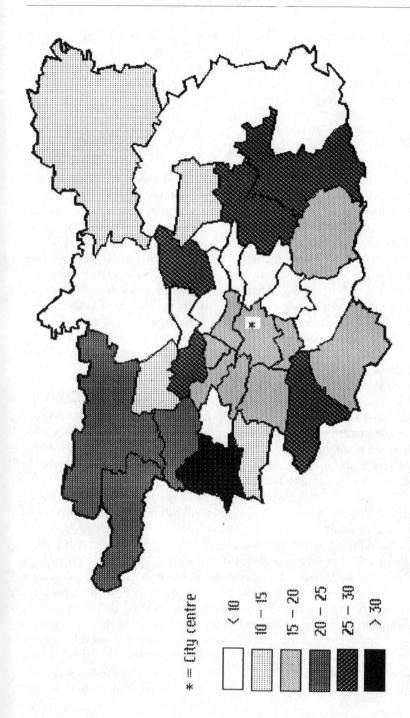

Figure 7.7 *Residence of 500 new clerical service workers in the City centre*

* = City centre

< 10
10 – 15
15 – 20
20 – 25
25 – 30
> 30

and actually simulate the impact of 500 new clerical jobs. Figure 7.7, for instance, shows that more of these jobs are likely to be taken up by residents of the north-western sector of the city than by residents of the deprived Urban Development Area: this is a point of obvious importance to urban policy formulation.

Retailing

For many years the retail model has been the basis of experimentation and illustration in a wide range of theoretical areas (see Birkin *et al.*, 1986). The retail model remains an important focus for model development, and recent work has taken three major directions. The first is concerned with the testing of theoretical ideas on patterns of equilibrium and dynamics in the real world retail environment of Leeds. This work draws upon the picture of retail change in Leeds since 1960 (see G.P. Clarke, 1986a). The main conclusions of this research are that it is possible to calibrate and fine-tune the equilibrium model to reproduce existing retail patterns and structures in a plausible way. Further, when the extension is made to dynamics (in this case using the retail grocery sector), actual system trajectories (or patterns of change) can be broadly reproduced and there is evidence that key parameters achieve critical values beyond which the nature of the structure of provision changes. (see G.P. Clarke, 1986a,b; G.P. Clarke and Wilson, 1986a,b).

The second approach to empirical work has involved extending the basic model into new areas of retail and marketing geography. These have included the motor industry, financial services and new models for out-of-town retailing (see Clarke and Wagstaffe, 1987). Out of this experience has emerged another order of magnitude of improvement in both understanding and usefulness.

The third area of study, and most directly relevant to public policy planning, has been our work on the recent influx of planning proposals for new shopping centres in Leeds. Although the main wave of supermarket, hypermarket and retail warehouses is probably over (although of course they continue to be built), retail planning continues to be faced with issues that cause great problems in terms of impact assessment. The latest in the long line of such trends is the new 'regional shopping centre', or less grandly new retail 'parks' based largely on comparison shopping. Over the last few years in Leeds there has been a large number of planning applications for such new developments. However, the impact assessments of such new centres are typically based on simple travel times around largely circular catchment areas. In particular it can be shown that catchment areas are not circular around individual centres and hence the impact on other centres is far more subtle. It is in these sorts of study that our work on performance indicators is also especially important, allowing the specification of residence-based welfare and accessibility indicators alongside centre-specific calculations on changing revenues.

Table 7.2   Summary indicators for the ten largest centres in West Yorkshire

Centre	SQ FT	CENREV	SPF	SCAT	CAT	CATCOMP
1 Leeds	3,075,000	576.0	187	682	844,492	27.73
2 Bradford	1,578,000	251.3	159	752	334,168	25.45
3 Huddersfield	602,000	85.1	141	727	116,621	28.33
4 Wakefield	577,000	75.7	131	761	99,518	25.58
5 Halifax	401,000	49.6	124	670	73,653	24.40
6 Keighley	307,000	34.4	112	758	45,415	31.01
7 Pontefract	227,000	20.7	91	690	30,010	20.14
8 Dewsbury	215,000	14.0	65	693	20,231	26.36
9 Skipton	224,000	14.5	65	795	18,244	39.41
10 Castleford	151,000	10.2	68	685	14,940	18.97

A simple example can be used to clarify ideas at this point. One of the many retail planning proposals for the West Yorkshire region is for the establishment of a large out-of-town centre in the Morley area. The first step in addressing the issues raised by this proposal is to use a performance indicator framework to appraise the existing situation. In Table 7.2., we present a list of indicators for major centres in the West Yorkshire region, relating to all non-food shopping activities. These basic indicators show that, as one would expect, Leeds is not only the largest centre in this region in terms of its physical size (measured as square feet of shopping space, or 'SQ FT' in the table), but also the largest in terms of centre revenue ('CENREV', in millions of pounds) and catchment population ('CAT'), defined here as the number of households using a centre as their only, or most frequent, destination for comparison goods shopping. Leeds also has the highest density of trading, measured in the retailers' sales per square foot ('SPF', in pounds per square foot). However, Skipton has the most affluent catchment area, with a high proportion of residents in the professional and managerial classes ('CATCOMP' is the percentage of the catchment population for which the heads of household are social class A or B), and the highest sales per catchment household ('SCAT', effectively the average expenditure of households in the catchment of the centre).

The fundamental question to ask now is 'what happens to this pattern if we add a new centre in Morley', and for the sake of argument we will assume that 600,000 square feet has been proposed. It is a straightforward matter to add this centre, and recompute the activity patterns and performance indicators, as shown in Table 7.3. Here we see that Morley generates a catchment of just less than 60,000 households, but this is only achieved with very large deflections from nearby Wakefield. This is, however, a rather sizeable development, and the rate of turnover is low relative to the large urban centres.

Table 7.3 Households in the catchment of the major non-food retail centres in West Yorkshire

| Centre | Catchment Populations | | Deflections | |
	Without Morley	With Morley	Gross	%
Bradford	334,168	161,503	8,020	2.4
Leeds	844,492	266,934	17,734	2.1
Huddersfield	116,621	99,022	5,831	5.0
Halifax	73,653	49,677	1,399	1.9
Wakefield	99,518	77,907	23,088	23.2
Morley	0	59,757	-	-

Having established the potential for revenue generation within a planned centre, and hence an evaluation of its profitability, another kind of question that the developer might wish to consider is what scale of development is most appropriate. This will be particularly pertinent if a mixed development of shopping with office space, leisure activities or even light industry is being contemplated. Again, this can be easily tested using the model. Table 7.4 shows a set of predicted revenues for a varying level of retail floorspace, showing a strong level of scale diseconomies. The implication of this result is that the new retail park has an obvious local catchment area which can be easily exploited, but a larger development will have to cut heavily into the catchment of the large and diverse competing centres of Leeds, Bradford, Wakefield, Halifax and Huddersfield.

For the planning authority which is responsible for the regulation of new developments, this information is all useful grist to the mill. However, additional insights can also be obtained by focusing on suites of residential performance indicators. One kind of indicator which is useful

Table 7.4 Performance of Morley with varying scales of development

Size	Turnover	Sales/ft	Catch pop	Sales/head
100,000	10,000	100	14,699	2,041
200,000	18,100	90	26,426	2,051
300,000	24,900	83	36,297	2,058
400,000	31,700	79	46,117	2,060
600,000	41,200	69	59,757	2,069

here is simply the average distance which consumers will travel for comparison goods. We present this, in Table 7.5 (in the column labelled 'Av. Dist.'), as an aggregate measure, although in a detailed study it might be necessary to break this down and look at the patterns for particular disadvantaged groups, such as the elderly. Table 7.5 also quantifies the 'effective delivery' ('Eff. Del.') of retail floorspace to residential zones in the neighbourhood of Morley. This is a notional measure of how much retail space is used exclusively by residents of a particular area, and hence can be a useful measure of the equity of a pattern of retail supply. Another way to look at this is as a provision ratio ('Prov. Ratio'), or effective delivery per head of population. Table 7.5 shows that there are considerable benefits to the local population from a new development in Morley, as one would expect, and as this area started as a relatively poorly provided one, this might be regarded as a significant (and quantifiable) plus by the planners.

Table 7.5 Residential performance in the catchment of Morley

| Post Code | Av Dist | Without Morley | | With Morley | | |
		Eff Del	Prov Ratio	Av Dist	Eff Del	Prov Ratio
LS27	7.2	2,343	20.5	5.5	3,300	29.0
LS10	7.0	2,443	19.8	6.0	3,050	25.8
LS11	5.0	2,426	20.5	4.9	2,828	23.8
LS12	6.7	2,665	24.1	5.6	3,148	28.4
BD 4	8.3	1,947	18.9	7.6	2,312	22.3
BD11	7.2	1,423	16.9	6.0	1,797	20.7
WF17	7.5	2,155	20.5	5.3	2,310	26.4
WF 3	6.7	2,012	23.8	5.9	2,198	25.9

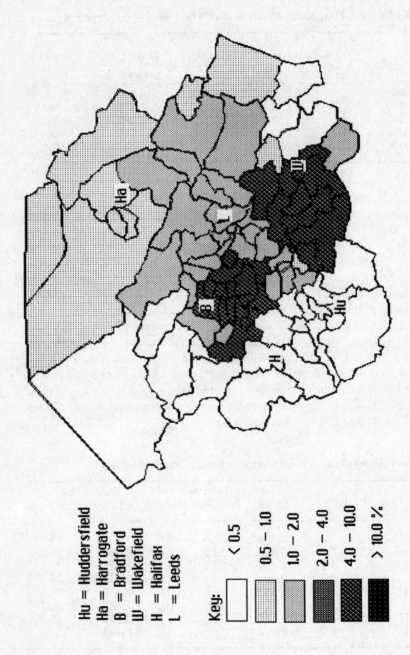

Hu = Huddersfield
Ha = Harrogate
B = Bradford
W = Wakefield
H = Halifax
L = Leeds

Key:

< 0.5

0.5 – 1.0

1.0 – 2.0

2.0 – 4.0

4.0 – 10.0

> 10.0 %

Figure 7.8a *XL Stores: baseline market penetration*

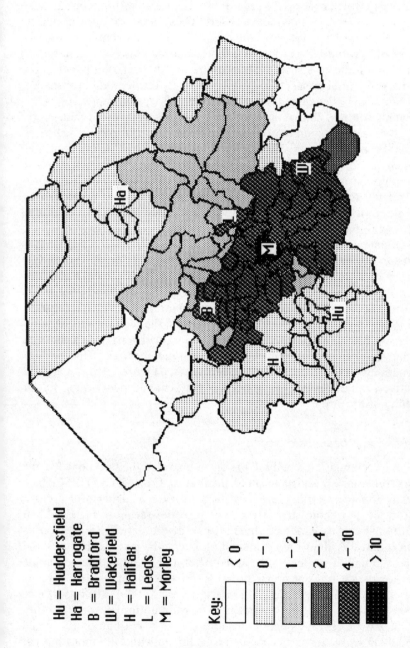

Hu = Huddersfield
Ha = Harrogate
B = Bradford
W = Wakefield
H = Halifax
L = Leeds
M = Morley

Key:

< 0
0 – 1
1 – 2
2 – 4
4 – 10
> 10

Figure 7.8b *XL Stores: market penetration with Morley shop*

It is also worth making the point that this kind of information can also be made available to individual retailers. These organisations will clearly be concerned to generate revenue estimates for a new centre in order to assess the profitability of locating there, but may also be interested in residential indicators such as the variation in their market penetration across residential zones. Many other kinds of residential performance indicator can be developed for either residential areas or supply zones i.e. shopping centres. One of the most interesting is 'group performance', or the comparison for an existing store of the known turnover against a model-based prediction. This is a very powerful tool in the evaluation of branch networks for all kinds of retailers. Note that when a new store or shopping centre is evaluated by simulation, this kind of facility also allows the retailer to assess the deflection pattern within his own organisation. This kind of analysis would typically be extremely difficult to quantify, but could have a crucial effect on decisions concerning marginal developments in the network.

As an example of this, let us consider an imaginary organisation, 'XL Stores', which sells mens' clothing. This multiple chain has three branches in West Yorkshire, one each in Leeds, Bradford and Wakefield. Figure 7.8a shows the pattern of market penetration which this implies. Although the three locations are the largest in the region, with spatially extensive catchment areas, this pattern of market penetration can be seen to be significantly concentrated around these centres. This pattern changes if we open a new branch in Morley (see Figure 7.8b). As this is a relatively small centre, this generates big peaks in penetration in the locality of Morley.

Housing and household incomes

The microsimulation data base has again provided the backbone for our work on housing and household attributes. From SYNTHESIS (see Birkin and M. Clarke, 1988), we can provide a highly disaggregate picture of household structures which provides far more richness than traditional census variables. Indeed, one of the deficiencies of census data is that it is almost ten years out of date and a major element of the microsimulation strategy is the updating of household and demographic characteristics. A discussion of progress with this line of research appears elsewhere (see Duley *et al.*, 1988; M. Clarke *et al.*, 1990). The importance of this work in application terms is to provide up-to-date ward profiles which are of obvious benefits in all areas of our urban and regional modelling work. In particular, it can be linked with our labour market modelling to give new insights into journeys to work patterns in the 1990s.

One particularly important application of SYNTHESIS has been in the production of individual and household incomes. Birkin and M. Clarke (1989) show how it is possible to construct an income-generating module based on the industrial sector and occupation of the individual within the SYNTHESIS data set using the New Earnings Survey. It is important to appreciate that this kind of data is simply not available to the academic or private community at the present time. Traditional proxies for income, such as owner occupation, professional workers or car ownership levels, have been the best we could get. Hence the new work reported here is of immense importance. On the commercial side there are obvious advantages to using incomes data. Academics, too, would find income data of great use in studies which attempt to understand the relationship between activity variables (i.e. any service utilisation) and measures of wealth/deprivation.

As an example, in Figure 7.9a we show average earnings for male workers in Leeds wards (in 1981). These estimates are obtained by synthesising data on the socio-demographic composition of wards, and the industrial and occupational composition of the work-force, with detailed data on wages from the New Earnings Survey (Department of Employment, 1982) which is powerful but crucially lacking in any small area component. The technique is discussed in more detail by Birkin and M. Clarke (1989).

The work on income generation is also leading to interesting research on benefits and taxation in general. Since we now have income coupled to household structure and occupation it is possible to apply direct sets of rules to generate state benefits and taxation for each household in the SYNTHESIS data set. This is particularly pertinent given the radical reorganisation of local taxation with the onset of the community charge or 'poll tax'.

As well as being crucially important in its own right, the work on individual incomes also allows us to generate (within the microsimulation framework) household incomes. This information can then be combined with data from the Family Expenditure Survey or other surveys to provide estimates of household expenditure (see Figure 7.9b). This kind of information is clearly of great value to retailers, as we saw in the section above, but is usually only generated by proxy measures involving some kind of clustering or social area analysis.

Education

An important component of the urban and regional information system is education, and this is one substantive area where extensions of the

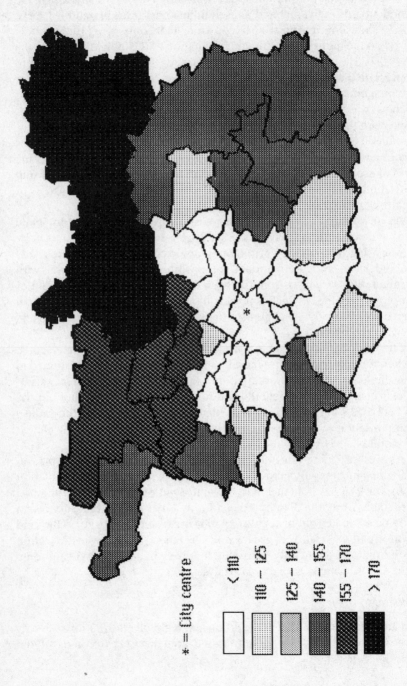

Figure 7.9a *Average male earnings, Leeds wards, 1981*

* = City centre

< 110

110 – 125

125 – 140

140 – 155

155 – 170

> 170

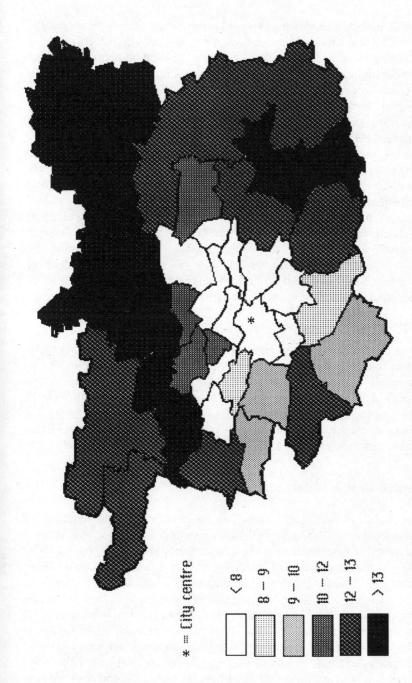

* = City centre

< 8

8 – 9

9 – 10

10 – 12

12 – 13

> 13

Figure 7.9b *Average household expenditure on fuel and power, Leeds wards, 1981*

modelling approach can be envisaged in the future. In many respects, the early 1990s have never been a more appropriate time to undertake such a modelling exercise in cities like Leeds, as the local planning authorities consider radical change. The problem of falling school rolls has been particularly significant in the development of plans for the complete reorganisation of primary and secondary education in the city. The most significant proposal is that of removing the current middle tier of education (9-13 year olds) to be replaced by a simple two tier system: (1) primary, 5-11 year olds, (2) secondary, 11-16. Clearly such radical reorganisation, coupled with the perceived need to remove the excess space caused by falling rolls, will result in some school closures and amalgamations. Such closures are bound to cause great resentment amongst parents and the public at large, and it will be interesting to report the results from our simulation exercises on the impacts of such changes. How will accessibility patterns be affected for example? Again, the performance indicator framework will be useful here in defining residence based and facility indicators. The same argument applies to other kinds of subsystem such as health, leisure and social services.

Comprehensive modelling

In our illustrations we have concentrated on static and comparative static applications of urban subsystem models. The possibilities are still more exciting if we can link these submodels together into a comprehensive whole with a dynamic focus. Thus when looking at a new shopping centre, for example, we are likely to be interested in the employment generated by such a scheme. How will this new employment affect the local labour market, and where is new housing to be found for these new employees?

Future research and developments

Three distinct areas of activity for future development need to be highlighted here. The first is in GIS development with the objective being to extend the functionality of the information systems to include a wider range of features such as overlaying and windowing. Of course these features (and many more) are contained within standard GIS packages, but equally these packages lack the kind of modelling capability which we have described above. In our case, GIS development is following model development rather leading: this is not the case in many other instances.

The second area for development is to conduct research into particular sectors: for example, many of the commercial applications of the work described above are in retailing, although in a very broad sense of that word. We have been working with clients in newspaper publishing and distribution, motor vehicle sales and retail breweries as well as high street and out-of-town retailing. The rapidly changing face of the retail environment (e.g. the demise of the high street and growth of new out-of-town centres; speciality retailing and niche marketing) must be paralleled directly in information systems design and, especially, modelling technology. The methods described in this paper are also being applied to a wider variety of sectors, including health, water resource and energy systems.

The third area where developments are planned is in expanding the 'local' dimension of our research — the local area is still the most appropriate point of reference for comprehensive modelling and information system design. The main requirement here seems to be to make local institutions aware of the power and applicability of urban and regional information systems. The institutions concerned might be local planning departments or Chambers of Commerce at the city level, or medium-sized to large businesses with regional interests.

One of the characteristics of this ongoing research programme is the diversity of its funding, being sourced as it is by a variety of private and public sector organisations. A second feature which is worthy of comment is that although the local area is still of interest, much of the research programme is focused on national, rather than regional issues. As most organisations are nationally oriented, this greatly enhances the number and variety of potential applications of a methodology which is fundamentally spatial rather than regional.

Bibliography

Birkin, M. (1990), 'An extended comprehensive model of the urban economy', in L. Anselin and M. Madden (eds), *New directions in regional anlaysis: integrated and multiregional approaches,* Belhaven, London.

Birkin, M. and Clarke, G.P. (1987,) *Synthetic data generation and the evaluation of urban performance: a labour market example,* Working Paper 502, School of Geography, University of Leeds.

Birkin, M. and Clarke, M. (1988), 'SYNTHESIS: a synthetic spatial information system for urban and regional analysis, with methods and examples', *Environment and Planning A,* 20, pp.1645-71.

Birkin, M. and Clarke, M. (1989), 'The generation of individual and household incomes at the small area level using SYNTHESIS', *Regional Studies,* 23(6), pp.535-48.

Birkin, M., Clarke, M. and Wilson, A.G. (1986), 'Theoretical properties of dynamic retail models', in J.H.P. Paelinck and A.C.F. Vorst (eds), *Shopping models,* Department of Theoretical and Spatial Economics, Erasmus University, Rotterdam.

Clarke, G.P. (1986a), *Modelling retail centre size and location,* Working Paper 482, School of Geography, University of Leeds.

Clarke, G.P. (1986b), *Modelling structural change in the retail grocery sector,* Working Paper 484, School of Geography, University of Leeds.

Clarke, G.P. and Wilson, A.G. (1986a), 'Performance indicators and model-based planning I: the indicator movement and the possibilities for urban planning', *Sistemi Urbani,* 1, pp.79-127.

Clarke, G.P. and Wilson, A.G. (1986b), 'Performance indicators and model-based planning II: model-based approaches', *Sistemi Urbani,* 2/3, pp.137-65.

Clarke, M. and Wagstaffe, S. (1987), 'Spatial analysis and geodemographics of financial services', Paper prepared for the Fifth European Colloquium on Quantitative Methods, Bardoneccia, Italy.

Clarke, M., Duley, C. and Rees, P.H. (1990), 'Microsimulation models for updating household and individual populations at the small area level', *Swedish Housing Journal,* (forthcoming).

Department of Employment (DOE), (1982), *The New Earnings Survey, 1981,* HMSO, London.

Duley, C., Rees, P.H. and Clarke, M. (1988), 'A microsimulation model for updating households in small areas between Censuses', Paper presented to the Workshop on Multistate Demography: measurement, analysis and forecasting, Kerkebosch Castle, Zeist, The Netherlands.

Harris, B. and Wilson, A.G. (1978), 'Equilibrium values and dynamics of attractiveness values in production-constrained spatial interaction models', *Environment and Planning A,* 10, pp.372-88.

Office of Population Censuses and Surveys (OPCS), (1982), *Census 1981,* County Reports, Part 1, HMSO, London.

Thygesen, L. (1984), 'A national register-based statistical system and its implications for local government', Paper presented at the 14th IARUS Conference, Copenhagen.

Chapter 8
A Land Information System for the monitoring of land supply in the urban development of Hong Kong

Anthony Gar-On Yeh

Introduction

Many of the foundations of geographic information systems (GIS) and land information systems (LIS) were originally laid down in the early 1960s. The early development of GIS/LIS was slow and applications were limited, partly because of the lack of software and partly because the computer hardware needed to deliver tangible results in practical, often widely distributed, planning environments was very expensive. Since the early 1980s, there have been major breakthroughs in the cost, speed and data storage capacities of computer hardware and great advancements in GIS/LIS technology and software. Consequently, GIS/LIS is becoming more affordable and more user-orientated and is being increasingly used by urban and regional governments and utility companies particularly in North America (see Millette elsewhere in this volume). In recent years, GIS/LIS has emerged as the major growth area in the application of computer technology to urban planning and management activities: it is being increasingly used in land information retrieval, development control, mapping (Newton and Crawford, 1988; Zwart and Williamson, 1988), site selection (Dangermond, 1983;) and land suitability analysis (Lyle and Stutz, 1987; McDonald and Brown, 1984).

While large-scale GIS/LIS applications involving massive volumes of data have been mostly confined to mainframe computer systems, recent advancements in microcomputer technology — especially in speed, cost and data storage capacities — have now meant that there is an increasing role for microcomputers in GIS/LIS developments and applications (Marble and Amundson, 1988). A microcomputer is often used to extract a subset of the global GIS/LIS data base from the central mainframe computer and to process that data locally at a remote site, such as a district planning office (Yapp *et al.,* 1986). This can save time in data communications between the central and district offices and free the mainframe computer at the central office for other uses. Under certain circumstances, this arrangement can also act as a substitute for mini or mainframe computers in small communities and districts where the data base is manageable in a microcomputer application (Yeh, 1988).

Land, particularly in Hong Kong, is a limited resource: monitoring the supply of land and making sensible decisions about the use of land are critical activities in effective land management. Development decisions can be far more effectively formulated if local government and the private sector have up-to-date information on the location, the amount, the suitability and the price of available land at their disposal. A comprehensive, widely accessible and rigorously defined land information system such as this will also enable local government to coordinate more effectively public investment in infrastructure, to integrate current land-using activities better and to increase the future supply of land. In addition, an LIS will allow planners to monitor the impact of public policies and regulations on the amount and location of developable land for housing and industry and to monitor the amount of vacant and derelict land. Without the continuous monitoring of land supply, inaccurate or out-of-date estimates of the availability of developable land can lead to increased land prices if supply is not allowed to expand to meet demand (Bollens and Godschalk, 1987).

Given both the growth in computing power and the planning imperative of accurate land supply monitoring and analysis, it is important to examine how the growing array of technological tools can be effectively applied in the area of land supply monitoring. It is not surprising that there has been increasing interest in applying the analytical and retrieval capabilities of computerised LIS to monitor land supply for urban planning and management in areas such as Hong Kong which have experienced considerable development pressure (Home, 1984; Godschalk et al., 1986). While most of the earlier developments in this field have been orientated towards mainframe computers, more recent systems have increasingly incorporated microcomputers. In this chapter, I propose to examine the progress that has been made towards the development of computerised land information systems in Hong Kong.

Land supply monitoring and land information system development in Hong Kong

Hong Kong has a total land area of 1,071 square kilometres and, from a population of 800,000 in 1931, the population had grown to around 5.5

million in 1986. Rapid population growth has exerted considerable pressure on the provision of housing, transportation, community facilities, employment and recreation. All these activities need land which is a very scarce resource in the main urbanised areas of Hong Kong. This problem of acute population pressure is further exacerbated by the fact that most of the land in Hong Kong Island and Kowloon consists of steep slopes which are not suitable for development. Land reclamation, both from the sea and from the terracing of hill slopes, has been the most commonly used method of providing the land necessary for urban development. The pressure on land created by the large and growing population has led to a land-use pattern which relies heavily on high density development. As a result, Hong Kong is one of the most densely populated areas in the world. The overall density in 1986 was 5,100 people per square kilometre. However, in the most densely populated District Board area, Mong Kok in Kowloon, the density in 1986 was 139,000 persons per square kilometre. Despite the lack of land and natural resources, economic growth in Hong Kong has been spectacular. It is a leading manufacturing, commercial and financial centre in Asia.

Urban development in the past was mainly concentrated in the Kowloon peninsula and the northern part of Hong Kong Island, along Victoria Harbour. With the emergence and implementation of a new town programme in 1973, the population has been dispersed into the once rural areas of the New Territories (see Figure 8.1). It is anticipated that approximately fifty per cent of the population will be living in the New Territories when all the eight new towns — which are at various stages of development — are completed. The spatial pattern of urban development will again be changed in the 1990s. In 1984, a Territorial Development Strategy (TDS) was formulated to provide a comprehensive long-term strategy for land and transport development to sustain the present momentum of urban development in Hong Kong into the 1990s and beyond after the new towns in the New Territories are fully developed. The direction of development will revert back to the main urban area along Victoria Harbour in the form of land reclamation along the water front. Land reclamation from the sea has long been a common practice in Hong Kong in expanding the urban land supply (Hong Kong Government, 1985). In addition, a Land Development Corporation was created in 1988 to use a public and private sector partnership approach to redevelop some of the older districts in the main urban area.

With the limited amount of developable land and the rapid pace of economic and urban growth, there has emerged a clear need for a computerised system which can store, retrieve and analyse land information for urban planning and urban management in Hong Kong.

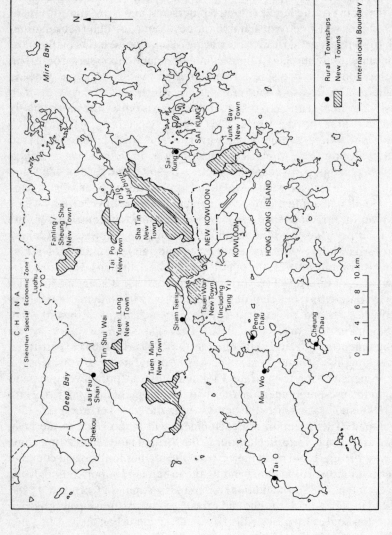

Figure 8.1 *New Towns in Hong Kong*

The system is also needed to monitor the pace and direction of urban development both spatially and sectorally. Furthermore, there are already in existence a number of well organised manual and semi-automatic data systems operated by various government departments which could be better utilised analytically, and better integrated to reduce data redundancy and improve quality, if they were transformed into an integrated geographically referenced data base. A land information system would not only increase administrative efficiency substantially but would also provide a powerful tool for planning and managing the scarce land resources of Hong Kong.

Land is both a vital factor of production and an asset for the generation of government revenue: in Hong Kong, the sale of land generally accounts for about twenty per cent of annual government revenue. With the boom in the economy in the mid-1970s, there emerged acute shortages of industrial and residential land and, because of dissatisfaction with past approaches for the production and sale of land (which paid insufficient attention to the needs of the private sector), a Special Committee on Land Production (SCLP) was established in May 1977. The main objectives of the Committee were to assess the amount of land that would be made available from the land development schemes already committed, to estimate the phasing of land availability, to identify potential development areas in which further land could be formed for development and to assess the amount of land, the costs, the planning implications and the development priority of land in potential development areas (Hong Kong Government, 1977, Appendix 2). As a result of this activity, the timing, location and quantity of different land-uses (especially for industrial, commercial and residential purposes) for the next ten years were compiled and this is reviewed on a year-by-year basis. By comparing the estimated land requirements and forecasting land supply for different land-uses, strategies for meeting projected shortfalls could then be formulated. For example, in 1981, the Committee identified a shortfall in the supply of residential land to meet future requirements. Consequently, consideration was then given to increasing the average plot ratio in the new towns which was at that time lower than that in the main urban area of Hong Kong (Hong Kong Government, 1981, pp.25-8).

The idea of a land information system (LIS) was proposed by the Town Planning Office of the then Public Works Department in the early 1970s. However, at the time, this idea was not given much attention. This was partly because of the failure to perceive the utility of the system by senior decision-makers and partly because of the limitations of the hardware, software and GIS technology which were available at that time. The idea was revived in the early 1980s when the Special Committee on Land Production recommended to the government the introduction of a computerised land inventory system which could provide the government, and possibly the private sector, with up-to-date land information for decision-making and planning (Hong Kong Government, 1981, p.41). At the time, the annual compilation of land supply data for the Committee was both time and resource consuming and an activity which would benefit substantially from the computerisation of land information. Apart from the need to produce statistics and to update maps more quickly, particularly in the case of Hong Kong where development is very rapid, savings in staff costs were found to be another important justification for the system. It is estimated that, once fully operational, the LIS will save over a hundred technical staff.

An *ad hoc* interdepartmental working group was set up by the then Director of Lands to conduct a review and subsequently to advise on the basic requirements of such a computerised land information system. In 1983, a team, headed by a Chief Land Surveyor, was set up in the Survey and Mapping Office of the Lands Department to carry out a more in-depth study of the potential of LIS. Although the LIS which was originally proposed was intended to serve all the government departments which either collected or used land information, it was found that it was difficult for different departments to agree on a common approach. Hence, it was found to be simpler to start the LIS as a project in the Lands Department which was, in effect, the government department which used land information most extensively. It was thought that once developed, the system could then be expanded to serve the needs of other departments.

The Hong Kong Land Information System

Origins of the Hong Kong LIS

The establishment of the Hong Kong LIS was approved at the end of 1986 and financial approval was given in mid-1987. In September 1987, a Land Information Centre (LIC) was set up in the Survey and Mapping Office of the Buildings and Lands Department and a project valued at HK$ 27 million (US$ 3.4 million) was put out to tender in January 1988. In May 1989, ARC/INFO with the support of a distributed network of SUN workstations was selected as the basic software and hardware for the Hong Kong LIS. A Steering Group composed mainly of representatives of the offices of the Buildings and Lands Department (the Town Planning

Office, the Land Administration Office, the Survey and Mapping Office and the Building Ordinance Office) was formed to guide the installation, operation and future development of the system. (As a result of the reorganisation within the Hong Kong government which took place in January 1990, the former Town Planning Office of the Buildings and Lands Department was upgraded to become a new Planning Department under a new Planning, Environment and Lands Branch, whereas the Land Administration Office, the Survey and Mapping Office and the Building Ordinance Office remained in the Buildings and Lands Department.)

Objectives and components of the Land Information System

The main objectives of Hong Kong LIS are to provide an efficient base for decision-making in land administration, to accelerate the updating and processing of land data and to establish a core system for other land-related systems. The system is composed of three application systems and one supporting system. The system will contain an integrated data base of both graphic and non-graphic data items. The main elements of the system are described below.

The Basic Mapping System (BMS) is used to produce maps and to maintain all planimetric details on the 1:1,000 survey sheets. Currently, there are approximately 1,700 active survey sheets. Maps are digitised using manual methods and subsequent updating will be carried out by digitising individual survey drawings or photogrammetric plots. The system can capture data directly from photogrammetric or ground surveys. Although the maps are digitised as separate sheets, they will be managed by a map librarian in the system which will make them appear as a continuous map to the user. All operations of the BMS, including data capture and subsequent updating, will be processed in the Land Information Centre which is the headquarters for the system's administration. In the past, it has taken between two and three years to update a survey sheet manually: with the LIS, survey sheets can now be continually updated and these maps can be reproduced at any time. The BMS will be extensively used by the Buildings and Lands Department and other Government departments, public utility companies and real estate agents for various urban management purposes.

The role of the Cadastral Information System (CIS) is to keep a record of land parcel boundaries and land status information for all private and government lots, to prepare cadastral plans and to provide enquiry facilities. It contains land parcel boundaries and associated information on land status, long and short-term leases, date of expiry of leases and

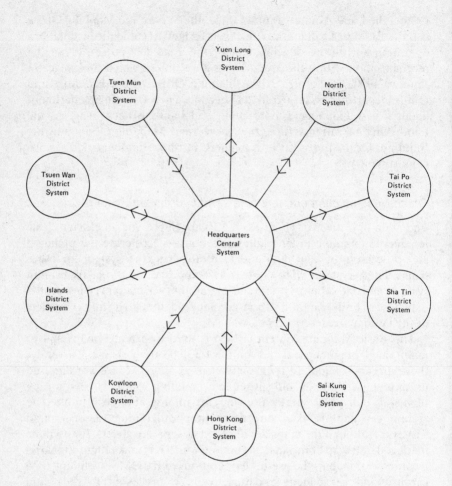

Figure 8.2 *District network of the Land Information System*

Source: Buildings and Lands Department, 1989

proposed land-uses. The CIS will be operated mostly in the ten District Survey Offices (see Figure 8.2) in order to provide timely and accurate land information and drawing services to their District Land Offices for land administration at the district level. It will also be used by other government departments as well as by the public. Land information at a district level is mainly used by the District Lands Office (DLO) which is responsible for the disposal of government land, for lease renewal and modification, for the control and management of government properties and for the acquisition of private land for public purposes and by the District Survey Offices (DSO). The DSOs are responsible for the definition and recording of cadastral boundaries, for the alienation of government land, for the allocation of land to various Government departments, for the resumption of private land and for the re-establishment of boundaries of old lots for redevelopment purposes. The total number of various types of land parcels and proposed land disposals amounts to around one million units. A summary of the respective volumes of land parcels by type is shown in Table 8.1.

Table 8.1   Hong Kong LIS: the volume of different types of land parcel

Types of land parcels	Number of land parcels
HK and Kowloon leasehold lot	35,000
Old Schedule lot	265,700
N.T. new grant lot	38,500
New town lot	1,900
Government land allocation	4,300
Temporary site	43,000
Proposed land disposal	2,600

Source: Hong Kong LIS, 1989

The role of the Town Planning Information System (TPIS) is to update and produce town plans, to keep site records, to maintain land-use and land supply information, to maintain facilities inventories, to allow interactive retrieval of graphic and textual data, to perform 3-D evaluations of the visual impact of proposed development and to perform data analysis and reporting (Chak, 1989; Yeh, 1990b). It will be operated by the system installed in the Town Planning Office. The system will be used by the different sections of the Town Planning Office, namely the Statutory Planning Section, the Planning Standards and Studies Section, the Sub-Regional Planning Section and the Central Information Unit. Data maintained in the TPIS can be extracted for analysis by other software packages such as AUTOCAD, Dbase III and SPSS-X.

The fourth element is the Geographic Information Retrieval System (GIRS) which is a supporting system to be used in parallel with the other three application systems. It is a tool to facilitate access to both graphic and non-graphic information in the system and consists of a number of standard programs which will enable users to define an area or a building using various identifiers and then enable them to select any combination of information from the BMS, the CIS and the TPIS modules in the system for analysis and further processing using other system functions. From a user point of view, this is the major interface with the land information system.

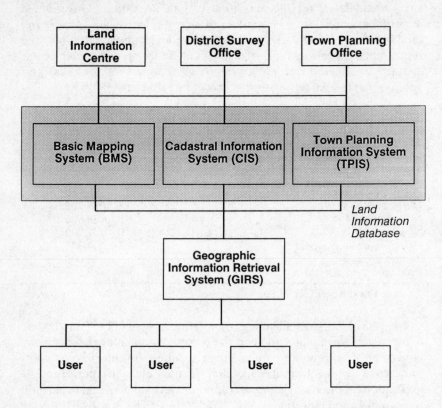

Figure 8.3 *Organisation and main components of the Land Information System*

The organisation and main components of the land information system are shown in Figure 8.3. The Land Information Centre, apart from maintaining the Basic Mapping System, is responsible for the management and operation of the whole system. The Cadastral Information System is operated and managed by the ten District Survey

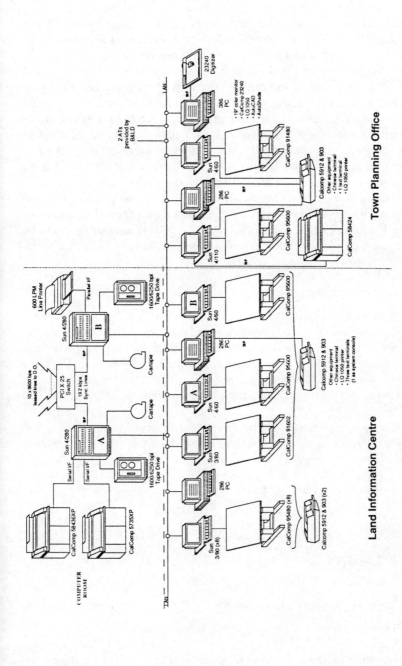

Figure 8.4 *The Hong Kong LIS: hardware configuration*

Source: Buildings and Lands Department, Jan 1990

Offices and the Town Planning Information System by the Town Planning Office. The Geographic Information Retrieval System enables users to access the information in the Land Information System.

The hardware is mainly located in the Land Information Centre, the Town Planning Office and ten District Survey Offices. The hardware configuration is comprised of SUN workstations, Calcomp electrostatic plotters and digitisers which are connected through an Ethernet local area network (this is shown in figure 8.4). Because of the frequent use of basic mapping information in each district, a copy of the full basic mapping data base of that district is kept in the District System. Updated information at the district level is transferred periodically to the Land Information Centre by means of tape cartridges. Initially, the system will be mainly used by offices in the Buildings and Lands Department and the Town Planning Office. When the system is fully operational, it will be opened up to users in other government departments and offices, and possibly to private firms at a charge.

The operation of the system

One of the basic features of the LIS is the 'layering' of geographic data. Map features are classified into groups and stored in separate layers in the computer. This can allow flexibility in data and map retrieval and in their combination. Initially, in the tendering exercise, there were 18 layers which covered basic features such as land parcels, Outline Zoning Plans, lamp posts and roads, though additional layers may be added in the final system.

All map features can be related to attributes through a common address referencing system. To standardise data, the following common address referencing system will be used:

a. House number and street name

b. Street intersection

c. Building name

d. Lot number

e. Coordinates

f. Tertiary planning unit number

g. District name

These identifiers can be used to access graphical or textual data.
The system has been designed to facilitate distributed processing and

this permits users to access the network while, at the same time, having processing power within their own workstation. This will reduce delays in response times that frequently occur in more centralised systems. Each workstation can also function almost independently from the other workstations in the network. Expansion of the system can be easily achieved by adding new workstations to the network. The land information system was set up in late 1989.

The process of digitising survey sheets and land parcel records will proceed district by district, starting with Kowloon, followed by Hong Kong Island and then moving into the New Territories. Once the data in a district are fully computerised, the system for that district can be put fully into operation. The total length of time that digitising the maps and entering data into the system is expected to take is around three and a half years — that is approximately four months for each district. The system should be fully operational by mid-1993.

The use of land supply information in the monitoring and programming of urban development in Hong Kong

The context for LIS use in Hong Kong

Monitoring the supply of land is important to ensure that future land supply will meet the social and economic needs generated by the urban development of Hong Kong. In Hong Kong, it is also used as a means of programming urban development. Since the early 1980s, land supply forecasts have been an important component in the preparation of development programmes for the new towns and for programming new reclamation and development areas in the urbanised areas and existing urban districts. Land supply in Hong Kong can be quite accurately estimated and effectively controlled because the government owns most of the land and most of the newly developed land is the result of government-sponsored land reclamation programmes.

Land in the new towns has been largely obtained from land reclamation: this has proved to be a cheaper and administratively easier option than acquiring privately owned agricultural land. The costly development of Fanling/Sheung Shui New Town, where land has largely been acquired from private owners, attests to this fact. Land reclamation is also the main method of urban development in the existing urban districts. To implement the common components of the two basic alternative patterns of growth identified in the Territorial Development Strategy (TDS) and to satisfy the projected needs of Hong Kong (Hong

Kong Government, 1985), about 550 hectares of land will be reclaimed along the waterfront of the main urban area. This will accommodate a proposed population of around 248,000 (Urban Area Development Office, 1988, p.1.02). There is very little vacant or undeveloped land within the existing urban districts and the vast majority of land zoned on plans for development has been taken up. Development in existing urban districts will be mainly in the form of redevelopment and land-use change within the existing built environment. However, some opportunities for new site formation do exist. These arise mainly from current squatter areas which are continuously being cleared (Pryor, 1984) and from new development sites on the fringes of the main urban area.

The planning and coordination of new town development is carried out by the new town development offices of the Territory Development Department (TDD), while the new reclamation and development areas in the urban area and existing urban districts are under the aegis of the Urban Area Development Office which is also under the Territory Development Department. The process of urban development, especially in the new towns, from new reclamation and from change of use sites in the urban area, commences with planning and engineering studies from which development plans and programmes are prepared. The implementation of the plans starts with the acquisition and clearance of land. It is followed by land reclamation, site formation and the provision of essential engineering services, such as roads, footpaths, cycle tracks, drainage, sewerage, water, gas, electricity and telephone cables. Land then becomes available for the building of public and private housing, factories, commercial buildings and community facilities. The government is actively involved in urban development through numerous buildings, engineering (civil and highways) and waterworks items in the Public Works Programme. The development of public sector housing estates is both funded and undertaken by the Housing Authority. The programming of all works is coordinated by the respective development office to ensure that, whenever possible and having regard to various constraints, the goal of balanced development both at the end of each stage of development and on the completion of the total programme is achieved.

The forecasting of the timing of land availability is one of the most important tasks in the preparation of the development programmes which are produced annually for the new towns, new reclamation and development areas in the urban area and in the urban districts in Hong Kong. The development programme, covering a period of ten years and rolling forward one year in each annual revision, is the main instrument used in the coordination of district planning and engineering works (Pun,

1984). Each updating takes into account the progress already made and the anticipated availability of government resources — both staff and money. The development programme is prepared based on some known factors, such as the anticipated funding for all TDD programmes and the Housing Authority's ten-year Housing Programme and Long-Term Housing Strategy. Information is gathered and compiled annually from various government departments on the timing of land reclamation, clearance, resumption, servicing and engineering projects in order to estimate the timing of land availability of different planned land-uses in the ten-year programme period. Based on this information, the timing of land availability, services provision and housing construction for the next ten years, commencing from the year when a programme is prepared, can be estimated.

In planning, it is important to be able to forecast likely population growth and distribution as this enables the planners to estimate the need for community facilities and infrastructure and to decide on their completion dates in order to support population growth. Working back from these dates, the programme of administration, design and construction works which must be undertaken in order to finish the projects on time to meet the needs of the growing population, together with estimates of financial expenditures, can be derived. Population forecasting by planning units is also an important component in the design of development programmes. Land supply information is used to forecast the likely population growth in the programme period.

Once finalised and approved, the development programme becomes the guiding document for all the agencies concerned in district development (Pun, 1984). It determines where and when public and private development can start. The urban planner uses it as a basis for planning and development decisions and the offices responsible for various facilities and services compile or adjust their own works, training and other programmes accordingly. The development programme is used to coordinate all the departments concerned to ensure that everything will be made available on time. It also serves as a guide to central policy decision-making, including the allocation of financial resources, and the preparation of territorial development strategies and sub-regional structure plans.

Land supply forecasts

Information on the timing of land reclamation, clearance and servicing and engineering projects is gathered from various government departments concerned with estimating the timing of land availability for different planned land-uses in a ten-year period. The serviced land available for development is mainly obtained from land formation and service engineering projects, from land resumption and clearance and from the expiry of short-term tenancies or temporary use.

Population forecasts

Urban development in Hong Kong, especially new town development, is mainly public sector housing-led (Yeh and Fong, 1984; Wang and Yeh, 1987): in the large new towns, over 50 per cent of the planned population is in public sector housing, and in the smaller new towns over 40 per cent is in public sector housing. As there is great demand for housing, particularly public sector housing, housing is often fully occupied soon after it is built. Future population increase in a district can be estimated accurately from the estimation of future housing stock based on land availability. A 'Flat Supply Method' which relates housing stock to population forecasts is used in the making of population projections for small areas in Hong Kong (Chak, 1986).

The general methodology is:

$$P_t = (H_{t-1} - D_t + N_t) \times PPOF_t$$

where P_t = forecast population in year t
    H_{t-1} = estimated total number of flats in year $t-1$
    D_t = number of flats to be demolished in year t
    N_t = number of new flats to be completed in year t
    $PPOF_t$ = persons per occupied flat ratio in year t

Based on the timing of land availability for different planned land-uses, population in different house types for the next ten-year period can be estimated. The community facilities required to support the population increase are then forecast in accordance with the Hong Kong Planning Standards and Guidelines. This is a very mechanistic but effective approach given the nature of the housing situation in Hong Kong.

Pre-LIS monitoring of land supply

The task of forecasting land availability is a time-consuming and labour-intensive process which has traditionally been done using semi-automatic methods and generally as an annual exercise. After the relevant information is collected, the timing of land availability of different parcels of land is marked on an Outline Development Plan or Layout Plan. The area of each parcel is calculated manually and recorded into a table by planning area or street block. Currently, the information on parcel area, together with other relevant land information, is stored in a microcomputer data base using Dbase III and a computer model for district population forecasting based on the above 'Flat Supply Method' is used. However, these are textual data which are not linked to computerised maps. The process of land supply estimation and population forecasting is tediously repeated annually in updating the development programme. Much time-consuming retabulation will be necessary if the spatial units for the estimation of land availability other than planning areas, such as sub-districts of district board, are required. A fully automated land information system with mapping capabilities will help to speed up the annual updating of the timing of land availability and the retabulation for different and varying spatial units. It can also provide the mapping of land availability and population distribution at any point of time to monitor development and assist decision-making. The system will allow the easy updating and querying of land supply data with the capability of mapping the output generated. Change in site boundaries can be easily updated without the need to recalculate manually net and gross site areas. It will also allow government departments and private developers to search for the availability of a particular type of land throughout Hong Kong. The system, if made accessible to the public, can help the private sector to make investment decisions, particularly given that a public-private partnership approach has been much emphasised in the urban development of Hong Kong.

The use of the Land Information System in the monitoring and programming of land supply and urban development

An LIS application in Sha Tin New Town

The need to monitor land supply was the main impetus in setting up the Hong Kong LIS. It was first initiated by the Special Committee on Land Production in the early 1980s because the process of preparing the annual estimates of future land supply for the Committee was both a time-consuming and a resource-consuming activity. The estimation of future land supply is also an important component in the preparation of development programmes which are updated annually for the planning and coordination of development in the new towns, new reclamation areas and development areas in the urban area and in the existing urban districts.

A study was carried out to examine the applicability and usefulness of a microcomputer-based LIS for the estimation of land availability and population for the new town development programme using PC ARC/ INFO (Yeh, 1990a). Although the study was carried out in Sha Tin, one of the new towns in Hong Kong, the methodology can be extended to the new reclamation and development areas in the urban area and urban districts. The LIS is divided into two main components which are shown

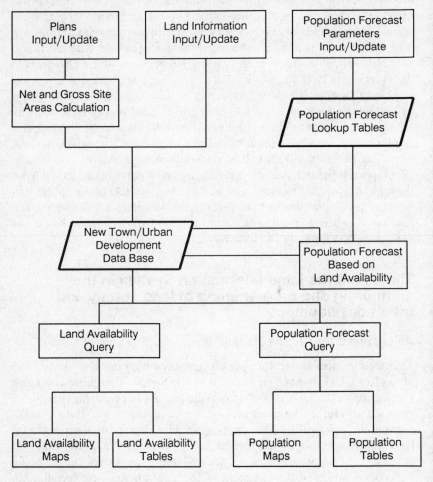

Figure 8.5 *Land Information System for monitoring and forecasting of land availability and population*

in Figure 8.5. The input component consists of modules for inputting and updating the new town development data base and population forecast look-up tables. The output component consists of modules for making land availability and population forecast queries. Both maps and tables can be output from the system.

PLANS INPUT AND UPDATE

Layout plans at a scale of 1:1,000 and outline development plans at a scale of 1:5,000 are digitised as separate coverages and later joined together to form one large coverage. The site number is used as the parcel identifier (PID) for relating the land information.

NET AND GROSS SITE AREA CALCULATION

Net site areas are automatically calculated by the LIS software. A small program is used to calculate the gross site areas required from the system. Gross site area is net site area plus a portion of the adjoining land, such as cut slopes, local roads and local open space. It is calculated by grouping sites into 'zone parcels' and apportioning part of the gross area of the 'zone parcel' according to the net area of the site. The gross site areas calculation routine is activated whenever a plan is updated. Gross areas, as well as net site areas, are calculated because they are needed to forecast population for different residential zones. In forecasting population for public sector housing zones, gross site areas are used while net site areas are used for population forecasting in private sector residential zones.

LAND INFORMATION INPUT AND UPDATE

The following information is collected and updated for each site:

1. planning area number

2. site number

3. layout plan number/Outline Development Plan number

4. statutory land-use zone

5. current land-use

6. land production mode and land status

7. public works programme item number and priority category

8. land formation date

9.   land serviced date

10. sale date

11. expiry date of short-term tenancy or temporary use

12. land availability date

13. number of living quarters (if known)

14. date of updating

POPULATION FORECAST PARAMETER INPUT AND UPDATE

The following parameters in the population forecast parameter look-up tables are input and updated:

1.   persons per occupied flat ratio (PPOF) by residential zone by year

2.   large site reduction factor (LSRF) by residential zone

3.   plot ratio (PR) by residential zone

4.   average flat size (AFS) by residential zone

5.   occupancy rate (OR) by year of completion

6.   domestic use rate (DUR) by residential zone

POPULATION FORECASTS BASED ON LAND AVAILABILITY

The flat supply method is used for estimating the population of different residential zones. Generally, residential land is considered to be occupied by residents three years after it has become available for development. Users can override the automatic calculation of year of occupancy and the estimated number of flats (living quarters) by inputting their own data perhaps based on more recent assumptions or to test varying scenarios. The basic formulas used in the system are:

a.   Existing residential land use and future residential land use with a known number of flats:

$$Pit = Fit \times PPOFit$$

b.   Future residential land use with an unknown number of flats:

  i.   Public Housing

$$Pit = GSAit \times PHPDt$$

  ii. Private Housing

$$Pit = NSAit \times LSRFi \times PRi/AFSi \times DURi \times ORj \times PPOFit$$

where Pit = population of zone i in year t
Fit = number of flats in zone i in year t
$PPOFit$ = person/occupied flat for zone i in year t
$GSAit$ = gross site area for zone i in year t
$NSAit$ = net site area for zone i in year t
$LRSFi$ = large site reduction factor for zone i
$PHPDt$ = public housing population density in year t
PRi = plot ratio for zone i
$AFSi$ = average flat size for zone i
$DURi$ = domestic use rate for zone i
ORj = occupancy rate for jth year of completion

Residential zones are classified as RS, R1, R2, R3 and R4 where RS (Residential Special) denotes public sector housing. Within the classification system, R1 has the highest density and R4 has the lowest.

Land availability and population forecast queries

Maps and tables of land availability and the derived population forecasts by land-use by year can be displayed either on the monitor or produced as hard copy from the plotter and printer. While under the pre-LIS arrangements, the user was limited to the spatial entities which could be used, under the PC-based system, the user can overlay a boundary different from that of the planning area to extract land availability and population forecast information within non-standard planning areas.

Conclusions

The Land Information System in Hong Kong represents a major step by the Hong Kong Government in developing more efficient and effective tools for urban planning and management in a rapidly growing area. The system was initiated in response to the need to monitor land supply because of the scarcity of land, the resultant high land prices and the population pressure in Hong Kong. An LIS was seen to be needed to provide the public and private sectors with up-to-date information on the timing, the location, the quantity, the suitability and the availability of developable land for the planning of rapid urban and economic development in Hong Kong. The estimation of land supply is also an important component in the annual preparation of development programmes for the monitoring and programming of urban development in Hong Kong.

An LIS has been shown to be an invaluable tool for handling the process of estimating land availability and for forecasting population growth in the programming and monitoring of urban development in Hong Kong where developable land is scarce. The system allows easy updating and query of the urban development data base with the capability of mapping any output. Any change in site boundaries can be easily updated without the need for the manual recalculation of net and gross site areas which are an essential ingredient in the production of population forecasts at the small area level. The system can enable government departments and private developers to search for the availability of a particular type of land in the already developed parts of Hong Kong and in those areas where new development is taking place. The system, if made accessible to the public, will help the private sector to make more informed investment decisions, particularly when a public-private partnership approach is an integral part of the process of urban development in Hong Kong. The LIS will save much staff time even if it is used only once a year for the preparation of development programmes for the new towns, new reclamation and development areas in the urban area and the existing urban districts. The benefits and flexibility of the system will be more appreciated and the system will become more cost-effective as the frequency of its use increases and as the variety of applications grows.

The computerisation of land supply monitoring in Hong Kong has encountered some difficulties. Some of the existing manual methods have had to be altered to accommodate computerisation (Werle, 1984). For example, in the manual method, parcels of land belonging to the same development but separated by a road were assigned the same site number. This is a workable solution in the manual system because the areas are calculated, added and then input into the data base as one record. However, within the LIS such a site would have to be entered as two separate parcels with unique identifiers and descriptors. The implication of this is that any process of computerisation does not simply involve a mechanisation of existing procedures: rather, it requires the redesign of procedures and, in some cases, a change in the basic concepts which underpin such manual or semi-automatic systems.

The flat supply method used in population forecasting based on land availability may work well in Hong Kong because of the great demand for housing and because over half of the flats are provided by the public sector. However, the applicability of the method would have to be tested for other cities which have different demands for housing and where the housing provided is mostly from the private sector. While this simplistic approach may not be readily applicable in the developed world, such methods, and also PC-based technologies, are likely to be far more applicable and usable in the less developed world. The approach outlined is very unsophisticated, yet due to the prevailing circumstances in Hong Kong, the method works very well. In other localities, and certainly where housing is allocated more by market forces than by bureaucratic procedures, a more sophisticated approach would have to be developed.

Theoretically, the LIS can be updated at any time, but often the frequency of updating is dependent on how the system is being managed, on the resources available and the frequency and degree of cooperation of other related departments in supplying updated information to the system. The LIS is also only a tool to facilitate the retrieval and analysis of land information. The frequency of use of the system, the interpretation of the information produced and the analysis of that information still rests with the planners and decision-makers. The system will contribute most effectively to decision-making and planning only if the planners and decision-makers are encouraged to understand and appreciate both the potential uses of the system and its practical limitations.

Bibliography

Bollens, S.A. and Godschalk, D.R. (1987), 'Tracking land supply for growth management', *Journal of the American Planning Association*, 53(3), pp.315-27.

Chak, K.C. (1986), 'Population distribution forecast — an example of computer application', *Planning and Development*, 2 (2), pp.47-52.

Chak, K.C. (1989), 'Town planning information system — the geographic information system for planners', *Planning and Development*, 5(2), pp.51-6.

Dangermond, J. (1983), 'Selecting new town sites in the United States using regional data bases', in E. Teicholz and J.L. Berry(eds), *Computer graphics and environmental planning*, Prentice Hall, Englewood Cliffs, NJ, pp.119-40.

Godschalk, D.R., Bollens, S.A., Hekman, J.S. and Miles, M.E. (1986), *Land supply monitoring: a guide for improving public and private urban development decision*, Oelgeschlager, Gunn and Hain in association with the Lincoln Institute of Land Policy, Boston.

Home, R. (1984), 'Information systems for development land monitoring', *Cities*, 1(6), pp.557-63.

Hong Kong Government (1977), *Report of the special committee on land production*, Government Printer, Hong Kong.

Hong Kong Government (1981), *Report of the special committee on land production*, Government Printer, Hong Kong.

Hong Kong Government (1985), *Planning for growth*, Government Printer, Hong Kong.

Lands Department (1984), *Town planning in Hong Kong*, Town Planning Division, Lands Department, Hong Kong.

Levine, J. and Landis, J.D. (1989), 'Geographic information systems for local planning', *Journal of the American Planning Association*, 55 (2), pp.209-20.

Lyle, J. and Stutz, F.P. (1987), 'Computerized land use suitability mapping', in W.J. Ripple (ed), *Geographic information systems for resource management: a compendium*, American Society for Photogrammetry and Remote Sensing and American Congress on Surveying and Mapping, Falls Church, VA, pp.66-76.

McDonald, G.T. and Brown, A.L. (1984), 'The land suitability approach to strategic land-use planning in urban fringe areas', *Landscape Planning*, 11, pp.125-50.

Marble, D.F. and Amundson, S.E. (1988), 'Microcomputer-based geographic information systems and their role in urban and regional planning', *Environment and Planning B*, Planning and Design, 15(3), pp.305-24.

Newton, P.W. and Crawford, J.R. (1988), 'Microcomputer-based geographic information and mapping systems', in P.W. Newton M.A.P. Taylor and R. Sharpe (eds), *Desktop planning: microcomputer applications for infrastructure and services planning and management*, Hargreen, Melbourne, pp.31-43.

Pryor, E.G. (1984), 'Squatting, land clearance and urban development in Hong Kong', *Land Use Policy*, 1(3), pp.225-42.

Pun, K.S. (1984), 'Urban planning in Hong Kong: its evolution since 1948', *Third World Planning Review*, 6(1), pp.61-78.

Urban Area Development Office (1988), *Urban area development programme: new reclamations/development areas — 1988 edition*, Urban Area Development Office, Territory Development Department, Hong Kong Government, Hong Kong.

Wang, L.H. and Yeh, A.G.O. (1987), 'Public housing-led new town development: Hong Kong and Singapore', *Third World Planning Review*, 9(1), pp.41-63.

Werle, James W. (1984), 'Problems in automating traditional land records data', *Computers, Environment and Urban System*, 9, pp.199-202.

Wiest, J.D. and Levy, F.K. (1977), *A management guide to PERT/CPM*, Prentice-Hall, Englewood Cliffs, NJ.

Yapp, G.A., Wilen, E.B., Gelinas, R.R. and Morrison, N.R. (1986), 'A microcomputer-based method for enhanced use of large land data systems in Canada', *Landscape and Urban Planning,* 13(3), pp.169-81.

Yeh, A.G.O. (1988), 'Microcomputers in urban planning: applications, constraints, and impacts', *Environment and Planning B,* Planning and Design, 15(3), pp.241-54.

Yeh, A.G.O. (1990a), 'Land information system for the programming and monitoring of new town development', *Environment and Planning B* (forthcoming).

Yeh, A.G.O. (1990b), 'Geographic information system for urban planning in Hong Kong', in United Nations Centre for Regional Development (ed)' *Geographic information systems applications for urban and regional planning: a state of the art review,* United Nations Centre for Regional Development (UNCRD), Nagoya.

Yeh, A.G.O. and Fong, P.K.W. (1984), 'Public sector housing and urban development in Hong Kong', *Third World Planning Review,* 6(1), pp.79-94.

Zwart, P.R. and Williamson, I.P. (1988), 'Parcel-based land information systems in planning', in P.W. Newton, M.A.P. Taylor and R. Sharpe (eds), *Desktop planning: microcomputer applications for infrastructure and services planning and management,* Hargreen, Melbourne, pp.44-53.

Chapter Nine
The application of a Geographic Information System for the allocation of land for housing in the Randstad Holland 1990-2015

S.C.M. Geertman and F.J. Toppen

The spatial policy context in the Netherlands

The role of spatial policy in the Netherlands is very prominent and can be explained by the fact that the Netherlands is a very densely populated country with considerable pressure on its land resources. The most densely populated part of the Netherlands is the area known as Randstad Holland, which has been a termed a 'rather peculiar horseshoe-shaped conurbation' (ter Heide, 1984, p.6). Its peculiarity arises from the fact that the economic, social and governmental functions often concentrated in the primate cities of other countries are distributed throughout the various cities and large towns which together comprise Randstad Holland. Within the Randstad, two development axes have been described (ter Heide, 1984): there is the 'northern wing' which includes Haarlem, Amsterdam, Hilversum, Utrecht and Amersfoort, while the 'southern wing' includes Leiden, The Hague, Delft, Rotterdam and Dordrecht. The agricultural area between these two wings is known as the 'Green Heart'.

Spatial planning in the Netherlands has its origins in the early twentieth century and according to Witsen (1977) is based on three principles: first, rather than being limited to simple land-use zoning, spatial policy in the Netherlands is orientated towards the positive guidance of spatial development; second, there is an awareness of the need to integrate spatial policy across various policy sectors; and, third, the development and implementation of spatial policy is seen as being the joint responsibility of national, provincial and municipal government, each of which is responsible for spatial policy at its own level. The three dimensions have major implications for the nature of the methodological issues and research problems which arise from the pursuit of spatial policy in the Netherlands: for example, the commitment to the positive guidance of physical development implies that there will be greater reliance placed on techniques for land-use allocation and optimisation than would be the case under other, less prescriptive, planning regimes. We propose to examine some of these methodological and research questions in this chapter.

The latest national-level spatial policy framework for the Netherlands was published by the Dutch government in 1988: this is the so-called *Fourth Memorandum on Physical Planning*. The Fourth Memorandum sketches a broad outline for the physical planning of the Netherlands until 2015. The National Physical Planning Agency was responsible for the preparation of the report and is currently involved in developing the plans for the implementation of the policies contained in the Fourth Memorandum. The intention is to use a geographic information system (GIS) to translate the Memorandum's broad policy statements into concrete location decisions: this reflects the first of the broad principles for physical planning listed above. In this chapter we elaborate on the way the Agency has embarked on the use of GIS for this purpose.

One important goal formulated in the Memorandum is the construction of more than one million dwellings in the already densely populated Randstad Holland. To this end, a user-friendly GIS application has been developed which will serve as a tool to allocate and evaluate new building sites in terms of criteria derived from the Fourth Memorandum. There are many lessons to be learned from this early application of GIS to solving practical planning problems. In particular, the pilot study described in this chapter indicates that the feasibility of using a GIS as a planning tool depends largely on the user-friendliness of the application. Of course, an appropriate application can technically be programmed, but its acceptance and use in practice will depend on the type of hardware and software available and on the characteristics of the users.

Although the use of GIS is expanding rapidly as a tool in physical planning, it has been found not always to meet the expectations and needs of users. Currently, only a few users are acquainted with all the detailed ins and outs of a GIS: this is not only with respect to its technical idiosyncrasies but also with the system's possibilities. Moreover, it is not clear whether special kinds of planning activities require a specific GIS solution or whether generic solutions can be developed.

The chapter contains three main sections. First, we discuss the physical planning context in the Netherlands and present in more detail the *Fourth Memorandum on Physical Planning*. We examine the importance of GIS for planning generally, and then discuss more specifically how GIS has been used by those involved in working out the plans to implement the Fourth Memorandum and we distinguish between different groups of

GIS users. In the second section, one specific use of GIS is illustrated in more detail. In a pilot study, a prototype GIS application was developed and tested. This prototype provides a system to support location decisions for the implementation of the urban and rural planning perspective, as indicated in the Fourth Memorandum. The application has been used as a tool to evaluate currently known potential building sites for residential use as well as to select additional sites. In the third section we highlight some of the more generally relevant experiences derived from our involvement with the practical application of GIS in a policy environment. In the conclusions we summarise how the application can serve the types of users identified and how the preferences, needs and expectations of the potential users can be met more effectively.

Physical planning in the Netherlands

Physical planning in the Netherlands has a long tradition. It emerged from legislation passed early in the twentieth century which was designed to regulate public housing. The 1901 Housing Act arose fundamentally as a reaction against the miserable living conditions in urban slum areas, but it also instituted the concept of the zoning of land for infrastructural development. Later on, in 1921, the spaces in between streets, squares and canals were also subjected to zoning regulations by local governments. In the following decades physical planning became increasingly independent of housing regulation and a separate body of planning and land-use legislation emerged.

In the Netherlands, land-use planning takes place at a hierarchy of levels: these are the national, provincial and municipal levels of government. Central government determines the main goals of physical planning policy, primarily by publishing policy documents (physical plans) and by supervising the policies of lower-tier authorities. The provinces draw up regional plans for parts of their territory and supervise the policy-making of the municipalities. A regional plan can be conceived as an outline plan, which serves as a basis for the development of municipal land-use policy. The municipalities in turn have an important dual role in the field of physical planning: they develop structure plans (local outline plans) for the entire territory and they also draw up development plans for parts of the municipality. The development plans are of great importance for physical planning because they are the only physical plans that are legally binding. This means, for instance, that building permits are normally issued only if they conform to the development plan (Meijer, 1986). Besides these national, provincial and municipal government activities, participation of the public in physical planning is prescribed by law. For instance, draft plans must be published and the public must be allowed to comment on them, either in writing, or verbally at a public hearing. The hierarchical and participative nature of Dutch physical planning is readily apparent from the above.

As we have indicated, central government intervention in physical planning dates from the beginning of this century. However, it was only in 1956 that the Dutch government published its first national plan (which was called *The West and the remainder of the Netherlands*). Two years later a more detailed document appeared (which was called *The Development of the West of the Country*) and two years after that (in 1960) *The (First) Report on Physical Planning in the Netherlands* was published. These documents sketched the main outline of physical planning policy up to the year 1980. This policy was aimed at slowing down the growth of the cities in the highly urbanised western part of the Netherlands (the Randstad Holland) and at the more balanced distribution of population and economic activity over the whole country.

Subsequently, in 1966 the central government published the *Second Report on Physical Planning in the Netherlands*, which outlined the proposed developments up to the year 2000. Again, special attention was devoted to a more balanced distribution of population and economic activity over the whole country. This was to be accomplished through the relocation of some government offices (main employment centres) outside the Randstad Holland and by stopping the process of suburbanisation, which would encroach upon the remaining greenbelt areas in the Randstad Holland. Within a national framework of decentralisation and population redistribution, ter Heide (1984, p.11) distinguished between what he termed the 'macrogeographic' and the 'microgeographic' policy implementation levels: at the macrogeographic level the policy emphasis was to decentralise population from the Randstad area in order to attain a more equitable population distribution throughout the Netherlands, but at the microgeographic level, the goal was to prevent suburbanisation by focusing development within city-regions by making more efficient use of already developed land — this was termed the principle of 'concentrated deconcentration'.

After 1970 however, it soon became clear that a balanced distribution of population and economic activity was a distant and unachievable ideal. Together with external factors such as economic depression, the oil crisis, the abrupt fall of the national birth rate and the First Report of the Club of Rome, this realisation led to the decision to publish a Third Memorandum, even though the Second Memorandum had only recently been published. The *Third Memorandum on Physical Planning* was published in three separate documents (an *Exploratory Report* in 1974, an *Urbanisation Policy Document* in 1976 and a '*Policy Document on Rural Areas*' in 1977), and contained a drastic change in approach to the dispersal policy. Because of factors like the rapid decline in population in the cities in the Randstad Holland, and the increasing growth in long-distance commuting, the stringent policy of dispersal was largely abandoned. Instead, the policy was aimed at reducing congestion in the Randstad area and improving the economic situation and the quality of life. The main goal of spatial policy contained in the *Orientation Report* (1975) of the Third Memorandum was to 'promote such spatial and ecological conditions that: (a) the true aspirations of individuals and social groups are realised as much as possible (and) (b) the diversity, cohesion and stability of the natural environment are guaranteed as well as possible'. These broad aims were to be implemented through active urban renewal policy and by designating growth centres at a short distance from the major cities (Meijer, 1986).

Several changes have occurred throughout the eighties which have had a fundamental impact on the framing of national spatial policy: these include a rapidly increasing internationalisation of society (culminating in the unification of Europe in 1992), increased prosperity and affluence despite high unemployment rates, a large decrease in the average number of occupants per dwelling and a continuing growth in levels of personal mobility as evidenced by a large increase in the number of private cars. In order to deal effectively with these trends, and to incorporate them into the framework for physical planning, the Dutch government published the *Fourth Memorandum on Physical Planning* in 1988. The planning horizon for this document was to be the year 2015.

The hallmark of the Fourth Memorandum is its selectivity, in that the Memorandum does not describe the full scope of physical policy but instead concentrates on the parts of existing policy that require revision or adjustment. Special attention is given to the maintenance and reinforcement of the international economic position of the Netherlands: in particular, the Randstad Holland has been identified as a locality which offers good opportunities for accelerated economic growth and a high standard of living (Ministry of Housing, Physical Planning and Environment, 1988). Consequently, there has been a realisation that there is a need to make more effective use of the existing land and infrastructural resources of the Randstad and to identify new areas for economic development which, while continuing to bring about the enhancement of the economic position of the Randstad, will preserve and enhance the natural environment. The scale and the difficulty of this task require the integrated and complex modelling of spatial development options together with their visualisation in a decision-support environment: this is an area where GIS can make a unique and effective contribution.

The role of Geographic Information Systems in physical planning

The National Physical Planning Agency is currently working out the details of how to implement the *Fourth Memorandum on Physical Planning*. In doing so, they wish to use GIS to translate general policy goals into concrete location and land-use decisions. One of the most pressing and prominent of these general goals is the need to satisfy the increasing demand for new housing — a need that is exacerbated by the ongoing process of decreasing household size and a growing demand for higher levels of housing amenity. Figures for the Randstad area indicate that there will be a demand for about one million new dwellings over the period 1990-2015: this projected demand for housing need in the Randstad is about half the total projected demand for the whole of the Netherlands.

As mentioned above, several organisations are involved in physical planning activities in the Netherlands. Besides the central government (several ministries), regional and local governments, various economic institutions and the private sector are also involved. Together they will decide on and negotiate over plans for the physical structure of the country. Because of the variety of actors involved, their data and information needs are very diverse and in this decision-making environment, a general geographic information system like ARC/INFO or Deltamap cannot reasonably be expected to satisfy such a wide range of demands for data and information. What is needed is a more sensitive appreciation of decision-makers' needs based upon a typology of actual and potential GIS users if GIS is to be more effectively used in a decision-support mode of operation. Such a typology is presented in the Scholten and van der Vlugt chapter of this volume and is particularly appropriate here. In a policy environment concerned with the implementation of physical plans, four kinds of GIS users may be distinguished (see also Scholten and Padding, 1988; Geertman, 1989): these are (a) information specialists, (b) policy analysts, (c) political decision-makers or senior management and (d) the public, the private sector and special interest groups. In terms of influencing policy-making and decision-making the nature and the quality of the interaction between the first three groups will be critical.

In this case study, the client for the pilot project can clearly be seen to be located within group (b), the policy analysts. The group is responsible for adding significant value to the output of GIS specialists. Within this group, potential GIS users consist of people preparing and working out policy in a specific area of physical planning. Normally they are not well enough acquainted with either the basic concepts of GIS or the often difficult-to-learn GIS command languages to use the programs unassisted — though this is changing. It is only with the help of an information specialist and/or a user-friendly shell that they can obtain the information they need. The dialogue between policy analysts and information specialists — the professional users of GIS — can be seen as critical in an applied policy environment. A second interface is also critical: this is the dialogue between policy analysts and the political and senior managerial decision-makers in group (c). This latter group often has limited or no understanding of statistics, computing, GIS or spatial analysis and so its need for clear and concise decision-support tools will be substantial. In addition, the legal requirement in the Netherlands to encourage public participation has major implications for the availability of unambiguous information on public policy for public consumption.

Up to the present, the National Physical Planning Agency has made a major effort to meet the specific information needs of these types of users. For the first group — the information specialists — the GIS package ARC/INFO was installed in the Agency in 1985. Because of the differences between the first and second type of user, some general user-friendly menu-driven shells have been developed around the GIS package (ARCSHELL by ARC/INFO (ESRI) and SPAT (Spatial Analysis Tool) which was commissioned by the National Physical Planning Agency). The development of SPAT was geared primarily to building a bridge between the information specialist and the regional planner (Zevenbergen, 1989). SPAT is a menu-driven generic tool based on ARC/INFO and contains five main functions (information selection, classification, the analysis of locational and attribute information, map creation and display and file handling) though progress is being made on incorporating spatial modelling techniques. The third and fourth user groups do not have direct access to the GIS system, and in order to meet some of the information needs of these latter types of user, a geographic query system has been developed from which they can obtain geographic information in both tabular and cartographic form — this is known as the RIA system (Scholten and Meijer, 1988).

In the policy environment we are concerned with here (i.e. the development of policy and programmes to implement the goals put forward in the Fourth Memorandum), regional planners form part of the very large second group of potential GIS users. They cannot make independent use of a common GIS because of the complexity of the GIS command language. Even with the help of general menu-driven shells like ARCSHELL or SPAT, only a small number of them are capable of working independently with GIS. To remedy this situation, a user-friendly GIS application should be designed and implemented to correspond with the above-mentioned shells. With such an application, those who are not GIS information specialists would be capable of manipulating at least the elementary functions of a GIS to satisfy some of their specific information needs.

Using GIS in planning: a pilot study

The context for GIS development

At the National Physical Planning Agency, several working groups are currently engaged in translating the general goals formulated in the Fourth Memorandum into more detailed policy measures for implementation. One major issue which has emerged from the plan is how to allocate new building land for housing development in the Randstad Holland. This problem has raised the question of how GIS could assist the working group in this complex and large-scale allocation problem. The GIS research group at the University of Utrecht was asked to design and implement a GIS application which would help the regional planners prepare and execute location decisions. Research at the GIS laboratory in Utrecht has emphasised the systematic development and evaluation of GIS applications, as well as the incorporation of more fundamental geographical methods like interaction and network analysis in GIS (Harts and Ottens, 1987; De Jong and Ritsema van Eck, 1989). Accordingly, the GIS research group initiated a pilot study.

In the first phase, an inventory was made of the needs of the people who were expected to use GIS, the suitability of the hardware and software and the existing data bases relevant to the research programme. The pilot study ended in February 1989 and resulted in two reports: one on the allocation of new building sites in nine urban agglomerations in the Randstad Holland (Toppen and Geertman, 1988) and a second on how to develop a special GIS application that would satisfy the needs of the working group responsible for making recommendations about housing allocation (Geertman and Toppen, 1989). The application based on the insights gained from the pilot study will be ready in April 1990.

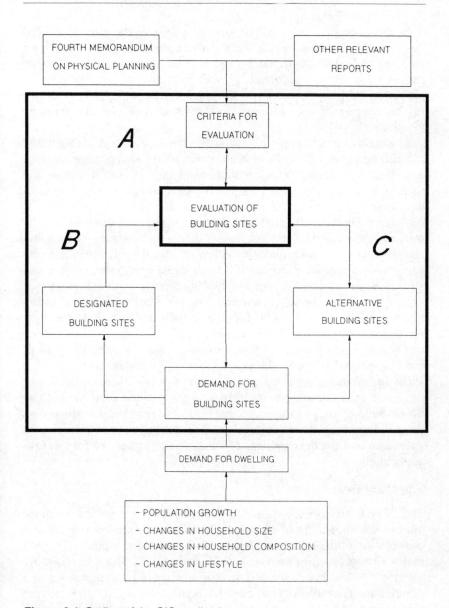

Figure 9.1 *Outline of the GIS application*

In this section we concentrate on the results of the pilot study and describe the problems which were encountered as well as the means selected of finding appropriate solutions. In addition, we give a brief review of a current follow-up project in which we are building an operational GIS-based model for goals-achievement analysis and design in the context of the selection and evaluation of possible building locations.

The main objective of the pilot study was to develop and test a method of building a prototype GIS application which would support the location and allocation decisions required in the implementation of the urban and rural planning perspective of the Fourth Memorandum. The process of developing and testing the method for building the GIS prototype is summarised in Figure 9.1. In short, the process proceeded as follows. The starting-point was the friction between supply and demand for housing: the projected demand for housing indicated a need to identify additional sites for housing development. Some of those new building sites had already been identified in various planning reports, both at the national and regional level, though, in addition to these sites, new locations within the Randstad area had to be found and then designated. In the next phase, the location of the already designated house-building sites and the newly designated house-building sites had to be evaluated in terms of criteria which had been derived from the *Fourth Memorandum on Physical Planning* and other pertinent reports: these criteria are described in detail later in the text as is the method used to evaluate house-building sites on these criteria. In the next section, we will discuss the problems encountered in developing the prototype and conclude with a discussion of the development of a user-friendly interface for this GIS application.

The study area

The *Fourth Memorandum on Physical Planning* stresses the strategic importance of Randstad Holland. So that the complex polynuclear conurbation of the Ranstad (Meijer, 1986) can compete effectively with major European cities such as Frankfurt, Brussels, Munich, Paris and London, it has been decided to concentrate future economic and population growth in this already highly urbanised part of the Netherlands (Ministerie van Volkshuisvesting en Ruimtelijke Ordening, 1988) — a task which will pose a major planning challenge. The pilot study was commissioned to evaluate proposed sites and to select new sites for residential development in this part of the Netherlands. From the Randstad area, which extends into the provinces of Noord-Holland,

Zuid-Holland, Utrecht and Flevoland, ten urban regions were selected for study (see Figure 9.2).

The Randstad possesses various characteristics which are critical to the

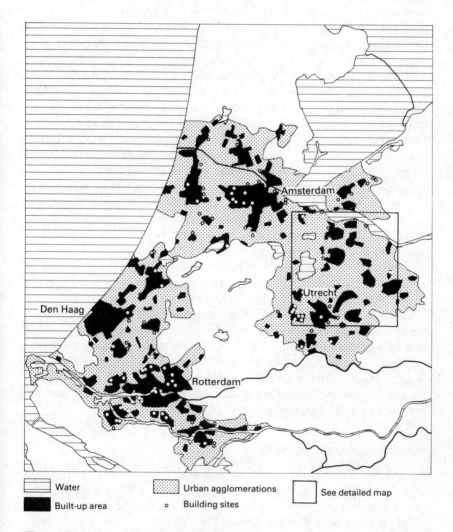

▭ Water	▦ Urban agglomerations	▢ See detailed map
◼ Built-up area	○ Building sites	

Figure 9.2 *Built-up area, urban agglomerations and designated building sites in the Ranstad Holland*

task of developing a GIS for urban policy analysis. First, the term Randstad Holland does not refer to a municipal entity but rather to a collection of urban agglomerations grouped around a rural core heart which despite its rural nature is considered an integral part of the Randstad area. Second, there is no clear hierarchical relationship between the cities in the Randstad because each city and large town performs a differing set of complementary urban functions: this complex pattern of overlapping urban fields creates a very difficult environment in which to make informed planning decisions.

Although to date all the physical plans state that no major urban growth should take place in the rural core of the Randstad area (the Green Heart), most of the population growth has taken place within the towns in this area. A process of rapid suburbanisation has also caused a substantial decline in population in the major cities and a sharp increase in commuter traffic: this has placed the Green Heart under increasing pressure. In order to avoid population growth all over the Green Heart, some towns have been designated as growth centres where further growth is to be concentrated. In addition, a continuing decline in the population of the four major cities was considered undesirable. Therefore, it was decided that growth should be directed to locations adjacent to existing major built-up areas. To reduce long-distance commuting, it was also decided that the growth of both population and employment should be accommodated as much as possible within the same urban agglomerations.

This, then is the policy environment and the spatial context in which the GIS application we are to describe was expected to play an important role in both the monitoring and the planning of the further growth and development of the Randstad. In addition, the GIS application was also expected to assist in the development and clear articulation of the general planning goals concerning the Randstad, as formulated in the various reports on physical planning.

Housing supply and demand

As indicated above, the Randstad will undergo significant population growth. The National Physical Planning Agency has used the 1987 population forecasts made by the Central Bureau of Statistics (CBS) to obtain a forecast of housing demand for various regions. The housing forecasts up to the year 2015 incorporated assumptions about the need to compensate for existing housing shortages, the replacement of older housing stock and lower occupancy rates. The growth in demand for housing will be the result of both population growth and of the substantial changes which are taking place in the size and composition of households. In addition, the CBS has recently updated its forecasts to reflect the fact that immigration into the Netherlands has been higher than expected.

These demographic pressures will create a demand for about two million additional dwellings in the Netherlands by the year 2015: about one and a half million of these additional housing units will be in the Randstad (this corresponds to an increase of about 20 per cent in the total housing stock). About a million units are expected to be built in newly developed sites and about half a million will be produced by the renovation and restructuring of existing sites in already built-up areas.

The pilot study was concerned only with the new building sites and, more specifically, only those sites within the ten selected urban agglomerations in the Randstad area. Given these constraints, the total demand specified for the study area for the period 1990-2015 was almost 600,000 dwellings. To satisfy this level of demand, a number of house-building sites have already been proposed for development in the period 1990-2015, but most of the supply of housing will occur in the period 1990-2000. These designated building sites were previously identified in the various physical plans of regional and local governments. In Figure 9.2, the designated house-building sites are indicated for the study area.

It is obvious that the designated building sites will not satisfy the total demand; the gap between supply and demand will be about 250,000 units. It is also obvious that the disequilibrium is larger for the years 2000-2015. This is partly due to the fact that this period is beyond the scope of the current regional and local plans in which those designated building sites have been identified. In order to reduce the imbalance between supply and demand in both periods, the National Physical Planning Agency has been seeking to identify alternative new house-building sites. In this context a user-friendly GIS is needed, not only for the interactive demarcation of new building sites, but also for the evaluation of sites which have already been designated.

Evaluation and selection criteria

Since this was a pilot study, the criteria were not determined as the result of an extensive selection procedure; rather, they were selected to test the feasibility of the method and to act as a baseline for the development of more sensitive and pertinent evaluation criteria. It should be stressed that in this pilot study the results of the evaluation exercise were not our main concern as we were most interested in the process of deriving the criteria themselves and testing the outcomes they produced when combined in different ways and with differing weightings. In subsequent research, and

in the final phase of the GIS application project, the selection of criteria
will be treated far more rigorously and extensively (Geertman and
Toppen, 1989).

For the pilot study, the selected criteria were derived from the general

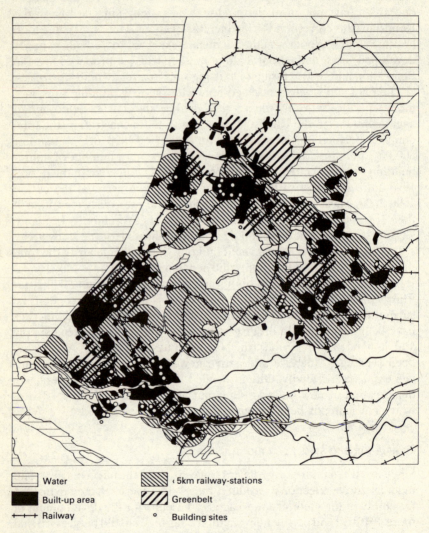

Water		‹5km railway-stations	
Built-up area		Greenbelt	
Railway		Building sites	

Figure 9.3 *Greenbelts and distance (<5km) to railway stations*

goals stated both in the Fourth Memorandum and in various other reports concerned with the spatial aspects of economic, housing, environmental, recreational and rural issues. The selection and the translation of general goals into criteria and appropriate GIS map layers is, of course, influenced by the presence, quality and accessibility of data bases. Therefore, the selection of criteria was partly dependent on the availability of the corresponding data bases.

Two of the general goals formulated in the Fourth Memorandum are of special interest in this context. First, in order to reduce mobility, the government has emphasised as desirable that the allocation of land for housing development should take place in the same urban agglomerations as the allocation of land for economic development. This measure has been designed to promote the use of public transport within the daily living environment. Therefore, the location of residential building sites had to be related to the location of employment, the existing built-up area and existing service and recreational facilities. The following criteria were derived from the general goal to reduce commuting (see Figure 9.3):

1. new building sites must be adjacent to existing built-up areas;

2. new building sites must be within a certain distance from secondary schools (these were both set at less than 3 km and less than 5 km);

3. new building sites must be located within a certain distance from an existing railway station (these were also set at both less than 3 km and less than 5 km).

The second general goal pertinent to our study was not only to create favourable conditions for potential house-building sites, but also to place conservation-orientated restrictions on such locations. These restrictions concerned the safeguarding of existing environmentally sensitive areas and protecting the Green Heart from large-scale construction activity. Therefore, potential house-building sites had to be evaluated in terms of the location and quality of agricultural areas, environmentally sensitive areas, large-scale open space and the greenbelt between built-up areas. In addition, environmental issues such as pollution and traffic noise had to be taken into account. Criteria derived from this second general environmental goal include:

1. no building sites were to be allowed within the greenbelt (i.e. the zones which are supposed to separate urban agglomerations — see Figure 9.3);

2. no building activity was to be allowed in large-scale open space (the scale factor having been translated into a measure of open space for each 100 hectare grid);

3. no building sites were to be allowed in environmentally sensitive areas (the importance of such areas was given as the number of hectares of natural landscape for each 25 hectare grid).

The data bases used in the study

The National Physical Planning Agency has a department which is resposible for collecting, updating and managing an extended set of data bases (Department of Information Supply). Some of the data bases which have been used in this pilot study are described briefy here — though in some cases in the pilot study, the data has been input especially for the study (such as for the location of railway stations). Four major information sources have been used in the study. The first was a data base which contains information about public services (such as medical centres, schools and sports facilities). The data here is available both for individual addresses and in a 25-hectare grid. This data base was used to derive the location of secondary schools in the pilot exercise. The second major source was the 'key planning decision maps' (*Planologische Kernbeslissing* — PKB) which are an integral part of the latest national physical plan. Information on the location of greenbelt and built-up areas was derived from these maps. The third source was the Physiognomic Landscape Information System, which contains a wealth of information about landscape and the natural environment using a 100-hectare grid system (this was used to identify large-scale open space in the pilot study); and the fourth source was a land-use statistics data base compiled from 1:10,000 topographic map sheets maintained by municipalities which was used to identify environmentally sensitive areas in the pilot study.

The prototype GIS application

The GIS application that we are currently developing has to serve several functions: for example, already designated building sites have to be tested against the selected evaluation criteria and the system has to be capable of being interrogated to identify and demarcate additional building sites to meet the remaining demand for new sites. The interrogation function can be achieved in a user-friendly GIS application by incorporating on-screen digitising to facilitate site selection. Such an on-screen digitising, site selection and demarcation facility requires a topographical map to paint a background within which sites can be selected. Alternative building sites can then be digitised and evaluated against selected planning criteria.

The recursive process which is contained within the prototype GIS application is summarised within the emboldened box contained in Figure 9.1. The application consists of the following elements which are denoted as Phases A, B and C in Figure 9.1.

Phase A: Criteria are selected to structure the planning decision. In this prototype only a limited number of criteria have been selected though in the final application many different criteria will have to be dealt with. It is also essential that different criteria can be combined in any weighted permutation required.

Phase B: The evaluation of already designated building sites. Given that sites may have been included within earlier planning documents as acceptable simply because they have not been subjected to such a systematic evaluation as we propose in this prototype system, it is probable that some sites will fail the evaluation process. Thus, it will be necessary to update the demand for housing to account for the rejection of sites already included which have been found to contravene the chosen selection criteria.

Phase C: The selection and demarcation of new sites. To meet the outstanding demand which has resulted from the rejection of sites in Phase B, additional potential building sites will have to be first identified and then added to the list of sites by way of on-screen digitising. To achieve this task, the GIS can be interrogated to display a surface containing information on existing built-up areas, on infrastructural elements and on open space: this can be then used as a background to assist planners in the site selection process. The newly selected building sites are then evaluated against the planning criteria which had not been used to structure the selection background. The outcome of the evaluation may cause the location, areal extent, and form of the possible building sites to be adjusted so that they conform with the various planning criteria. In fact, this is an iterative process in which new building sites are digitised and adjusted until the total demand for housing is satisfied.

To illustrate this GIS application, some results of the case study in which the prototype was used are presented here. Figure 9.3 gives an example of a combination of elements from phases A and B in which criteria are selected and combined and subsequently used to evaluate building sites. The two criteria selected and combined were that house-building sites should be a maximum distance of 5 km from an existing railway station and that no building activity should be allowed in the greenbelt.

Although only a small number of criteria has been selected and used for illustrative purposes, it is interesting to consider the effects of this relatively unsophisticated evaluation process on the building sites which have been designated using more traditional planning methods. The results, which were given originally in areal units, have been translated into an estimated number of dwellings using a mean density assumption of 50 dwelling units per hectare. In Table 9.1 the results are presented for the evaluation of all designated building sites in the study area on the criteria of less than 5 km distance from a railway station and preservation of greenbelt.

Table 9.1 Consequences for housing supply of satisfying the criteria of sites being less than 5 km from a railway station and not within the greenbelt over the period 1990-2015

	CRITERIA		
	Criterion of less than 5 km from a railway station:	Criterion of greenbelt:	Combination of both criteria:
Number of dwellings allowed	302,700	404,000	279,100
Number of dwellings rejected	126,300	25,000	149,900
Number of potential dwellings on building sites	429,000	429,000	429,000

More than one-third of the areal extent of the designated building sites did not satisfy both of the selected criteria and were thus rejected. The criterion of less than 5 km distance to a railway station was more discriminating than the greenbelt criterion. Such a rejection of designated building sites indicates the need to demarcate new building sites in order to overcome the imbalance between supply and demand.

This example shows the superiority of GIS methods over more traditional planning techniques.

In Figure 9.4 we give an impression of the background against which

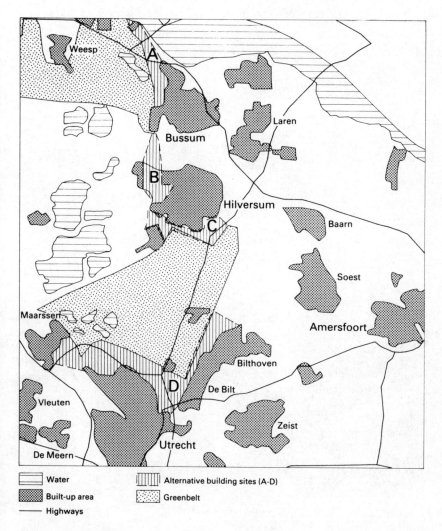

▭ Water	▥ Alternative building sites (A-D)
▦ Built-up area	░ Greenbelt
── Highways	

Figure 9.4 *Alternative building sites in relation to the built-up areas and greenbelts*

the selection and demarcation (on-screen digitising) of new potential sites was carried out. The process contains the following steps: first, the planner selects a part of the study area (this is shown in the inset in Figure 9.2); a background is then created on the screen which contains information on (in this case) built-up areas, highways, lakes and

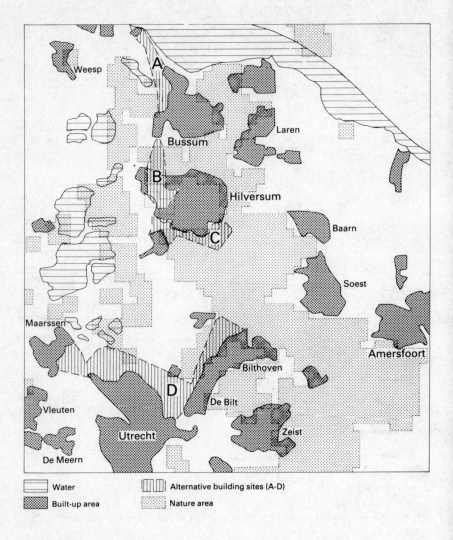

Figure 9.5 *Alternative building sites in relation to nature areas*

greenbelt; possible building sites are then drawn on the screen (these potential house-building sites are labelled A, B, C and D in Figures 9.4 and 9.5). Later, other criteria (such as environmentally sensitive areas or nature areas) are selected and used for the final evaluation of the possible building sites (this is illustrated in Figure 9.5). In Table 9.2 we indicate the amount of land in the possible building sites A, B, C and D which conflict with the nature area criterion.

From Table 9.2, it can be seen that only 15.1 per cent of the designated new building sites are in conflict with the criterion of nature areas. The results show that the prototype application can serve as an effective and flexible tool to select and evaluate house-building sites. However, the use of the prototype did expose some drawbacks and we discuss these in the next section.

Table 9.2 The amount of conflict between possible building sites and nature interests

Alternative building site *	Surface of design in ha	Amount of conflict in ha **	percentage
A	444	15	3.4%
B	550	145	26.4%
C	510	135	26.5%
D	2,095	248	11.8%
Total of A,B,C and D	3,598	543	15.1%

Notes: *   For the location of the alternative building sites A, B, C and D, see Figure 9.5.
  **   Total areal extent of nature areas in grids (of 25 hectares), even if they are only partially covered by an alternative building site.

Problems encountered in the development of the GIS pilot study

In the pilot study we have described above, the evaluation of existing building sites was comparatively easy. The main problems seemed to exist in handling the extensive data bases, in the task of on-screen digitising, in the capabilities of the available hardware configuration and in the user-friendliness of the software. Several of these problems are briefly described below.

Using different data bases

As indicated above, only a limited selection of criteria was made in the pilot study. Even then, different data bases with dissimilar data structures and scales were required in order to create map layers that could be used for the evaluation of house-building sites. Besides the more fundamental problem of combining data bases that use a grid structure as their basic areal unit and others that use a polygon structure, most of the data bases lacked a clear spatial reference. Therefore, various information specialists were needed to assist users in translating data bases into map layers appropriate to the nature of the required application. Another problem was the lack of documentation (i.e. meta-information) about the structure and contents of the data bases and, perhaps even more seriously, how to extract the required information from the data base. In the final application, users will be offered a large number of map layers prepared by information specialists. The specialists will base their selection of map layers on inquiries made among potential users, on the availability of the information and on the ability of that information to be used within a GIS.

User-friendliness

Most GIS are not user-friendly — even GIS specialists encounter problems with most software packages, but in general those individuals are able to deal with most of the problems they encounter. However, as Ottens (1990) stated: 'Planning involves much more than automated information processing (and) planners only seldom use information systems intensively and on a daily basis'. Planners are not experts in using information systems and should not have to spend large slices of their time navigating their way through the hardware and software jungle. As a planning tool, GIS must be efficient and problem-oriented; the functions may be few and must be easy to use.

The National Physical Planning Agency has already started work on the development of a user-friendly tool, SPAT: Spatial Analysis Tool (Zevenbergen, 1989). SPAT offers users the opportunity to select, classify, query and plot map layers and to perform some basic analytical functions using one or two map layers. Although SPAT offers easy access to the most frequently used GIS functions, some shortcomings have emerged in the use of the GIS prototype application. One of the major problems is the small number of map layers which can be dealt with at one time. This is especially serious because in this specific application, the user should be able to select and combine a large number of map layers with some flexibility as this is reflective of the complexity of the planning task he faces. Also, classifying and plotting is still cumbersome. What is lacking in ARC/INFO as well as in the SPAT shell is a tool for effective, efficient and simple on-screen digitising.

On-screen digitising

The ability easily to digitise new lines and/or polygons on-screen by using a mouse or cursor keys, to adjust the results and to translate them in ARC/INFO map layers (coverages) is of major importance in this prototype application. The results of on-screen digitising (i.e. the demarcation of new potential house-building sites) were far from precise, and adjusting the form, and/or the location, of the polygons was only possible by starting the whole process all over again: this is an area where improvements are required.

Tabular reporting

Besides graphic output, tabular reports were found to be essential in the evaluation of the building sites in relation to the planning criteria selected. To create a tabular report including only the appropriate amount of information was again found to be cumbersome, especially when dealing with complex maps (for example, those maps which are produced as a result of applying several overlays).

Hardware environment

Until now, GIS users at the National Physical Planning Agency have shared ARC/INFO on a mainframe. The same computer was also intensively used for other activities besides GIS. Access time was problematic and found to be unacceptably slow when planners wished to use the GIS application interactively: this would be a major constraint in a practical planning environment.

Ongoing research

In the current follow-up phase of research, scheduled to be completed in April 1990, the application will be made more user-friendly. The Spatial Analysis Tool (SPAT) will be adjusted and fine-tuned to the needs of this application. In addition, a larger number of data bases is being used to define an extended set of planning criteria. All the criteria will be made available to the users of the application in separate, easily accessible map layers. In this way planners should be able to select, combine and weight the criteria they need for evaluating and designing house-building sites within a user-friendly environment. A special ARC/INFO macro will also be developed to help the planners design alternative house-building sites. The program will support the digitising process and also the translation of the new potential house-building sites into map layers. Furthermore, a base map showing some general information and some basic criteria will be created in order to assist planners in the design process. New hardware (workstations) will hopefully allow users to access the application more effectively.

Potential users will be involved in all stages of the process of developing a more user-friendly application. Users will be asked to criticise the tools developed on a regular basis and they will also assist in the pre-selection of data bases. Later on, the developers and the users will work closely together in selecting the final set of variables to be incorporated within the application and in the preparation of the final map layers.

Conclusions

In this chapter we have attempted to achieve several objectives. First, we have described the spatial planning context in the Netherlands and also alluded to the importance of the Randstad Holland within the Netherlands space economy. Within the framework of this descriptive material, we then described how we set about the development of a GIS application to assist in the complex task of housing land allocation within a highly developed and environmentally sensitive polynuclear conurbation.

We have described several elements of the prototype GIS application in some detail and have illustrated these elements with maps and tabular output to give some tangible idea of the usefulness of the system in a practical planning and policy-analytical environment. A preliminary assessment of the pilot study indicates that it has been a successful application of a GIS within an applied planning environment both in terms of its contribution to more sensitive and systematic spatial planning and as a learning exercise which will contribute to the development of a more effective dialogue between information specialist and regional planners.

However, some critical remarks must be made about our prototype GIS. First, for non-specialists (i.e. the second group of GIS users we referred to earlier) using the prototype, it has proved difficult to select and combine criteria and to apply differing weights to the planning criteria selected. This is even more difficult when some flexibility within this process is needed. Second, on-screen digitising has been found to be a difficult task for non-specialists. The translation of the results of on-screen digitising into real computer maps is also time-consuming and runs contrary to the desire to digitise interactively, to evaluate options and to adjust house-building sites according to the criteria selected. Third, while tabular information is an essential part of the presentation of results in the GIS application, good tabular representation of an evaluation of house building sites requires extensive knowledge of the GIS software used.

A common GIS like ARC/INFO is flexible, integrated and efficient but it is by no means interactive (user-friendly), although ARCSHELL offers a real improvement. Furthermore, it is only partially effective: the analytical functions are limited particularly in their ability to be integrated with spatial models; the query function demands extensive knowledge of the GIS used; and the on-screen design and evaluation of polygons (building sites for new houses) is cumbersome. This has resulted in the request for the design and implementation — with the help of the Arc Macro Language (AML) — of specific, user-friendly and problem-oriented GIS applications. Of course, a data base much more appropriate for, and finely tuned to, the needs of planners and ultimately political decision-makers has yet to be built and in particular the application has to be made more user-friendly. In conclusion, we reiterate the importance of making GIS tools user-friendly and flexible because only then can GIS be useful in a practical physical planning environment.

Bibliography

De Jong, T. and Ritsema van Eck, J. (1989), 'GIS as a tool for human geographers: recent developments in the Netherlands', Paper presented at the Deltamap User Conference, Fort Collins, USA.

Geertman, S.C.M. (1989), 'The application of a geographic information system in a policy environment: allocating more than one million dwellings in the Randstad Holland in the period 1990-2015', ARC/INFO Fourth Annual ESRI European User Conference, Rome.

Geertman, S.C.M. and Toppen F.J. (1989), *Voorstel voor het Hoofdonderzoek Ruimte voor de Randstad, Ruimte voor de Randstad,* deel 2, Faculty of Geographical Sciences, University of Utrecht, Utrecht.

Harts, J.J. and Ottens, H.F.L. (1987), 'Geographic Information Systems in the Netherlands: application, research and development', Proceedings of the International Geographic Information Systems (IGIS) Symposium: *The research agenda,* volume III (Applications and Implementation), 15-18 November 1987, Arlington, Virginia, USA, pp.445-57.

Meijer, H. (1986), *Randstad Holland; Information and Documentation*

Center for the Geography of the Netherlands (IDG), Utrecht/The Hague (2nd revised edn, 1986).

Ministerie van Volkshuisvesting en Ruimtelijke Ordening (1988), *Vierde Nota over de Ruimtelijke Ordening, deel a: beleidsvoornemen,* Tweede Kamer, vergaderjaar 1987-1988, 20.490, nrs. 1 en 2, SDU, 'sGravenhage.

Ministry of Housing, Physical Planning and Environment (1988), *On the road to 2015; comprehensive summary of the fourth report on physical planning in the Netherlands: policy intention,* The Hague.

Ottens, H.F.L. (1990), 'Application of GIS in urban and regional planning', in H. Scholten and J. Stillwell (eds), *Geographical Information Systems and urban and regional planning,* Kluwer Academic Publishers, Dordrecht (forthcoming).

Scholten, H.J. and Meijer, E.N.(1988), 'From GIS to RIA: a user-friendly microcomputer orientated regional information system for bridging the gap between researcher and policy maker', Paper presented to the URSA-NET Conference, Patras, Greece.

Scholten, H. and Padding, P. (1988), 'Working with GIS in a policy environment', ARC/INFO Third Annual ESRI European User Conference, Kranzberg, W.Germany.

Ter Heide, H. (1984), 'Demographic research questions arising from spatial policy in the Netherlands', in H. ter Heide, and F.J. Willekens (eds), *Demographic research and spatial policy,* Academic Press, London.

Toppen, F.J.and Geertman S.C.M. (1988), *Ruimte voor een Miljoen Woningen in de Randstad? Ruimte voor de Randstad,* deel 1, Faculty of Geographical Sciences, University of Utrecht, Utrecht.

Witsen J. (1977), 'Crucial physical planning decisions', *Planning and Development in the Netherlands,* 9(2), pp.99-114.

Zevenbergen, M. (1989), 'Spatial analysis tool: making ARC/INFO available to the regional planner', Paper presented to the 29th European Congress of the Regional Science Association, Cambridge.

Chapter Ten
Small area information systems: problems and prospects
L.G. O'Brien

Introduction

The market for geographically-based information systems is booming. Government and business alike have come to realise that success in our 'information economy' depends on having access to reliable information for small geographical areas and being able to manipulate that information for profit. Whether the objective is to deliver statutory services or sell products, there is an increasing belief that 'geography matters' and that it cannot be pushed to the margins of decision-making as in the past. However, amid the euphoria there are growing doubts about the ability of these systems to live up to expectations. The main worries concern the quality of the information in the systems and how it may be used to affect peoples' lives. If good data can lead to good planning, dubious data must lead to dubious planning, unless the necessary checks and health warnings are applied. The most obvious of these health warnings is that small area information systems are not a universal panacea for planning problems. They offer a way of aiding planning, not replacing it altogether.

This chapter is concerned with appraising a specific small area information system: the National Online Manpower Information System (NOMIS) which has been described as Britain's largest and most successful GIS (Rhind and Mounsey, 1988). By most standards this is a successful system which has been developed over the years to meet many of the information needs of a variety of types of client. The nature of the services provided is described in the second section of this chapter. Some of the technical features which have led to the current system are outlined in the third section and assessed in the fourth section. The fifth section broadens the discussion by looking at some of the more general geographical concerns underlining the use and misuse of information systems.

The NOMIS system

Overview

The National Online Manpower Information System is a geographically-based data retrieval and spatial analysis system developed by the Department of Employment Training Agency in conjunction with the Geography Department at Durham University. It offers access to many millions of items of labour market and demographic information for small geographical areas in the UK (Townsend *et al.*, 1986; 1987). These include wards defined for the population census, wards based on Department of Employment definitions and job centre office areas defined by the Department of Employment as part of job search (see Table 10.1). Spatial aggregates may also be created from these building blocks which correspond to administrative, political and functional spatial units (see Table 10.2).

Table 10.1 Small geographical areas within NOMIS

Type of unit	Number of units in system
OPCS ward	10,903
DE ward	10,519
Job centre office areas	916
NHSCR areas	121

Note:	OPCS	— Office of Population Censuses and Surveys
	DE	— Department of Employment
	NHSCR	— National Health Service Central Register

Source: O'Brien *et al.,* 1988a

The primary purpose of NOMIS is to provide the Training Agency (a national government agency) with a flexible tool for accessing the most contemporary information on the economic performance of these small areas. Typical sources of this information are drawn from the nationally available sources of statistics in the UK and include the monthly unemployment count, the monthly and quarterly counts of job vacancies and other employment opportunities, quarterly sub-regional migration returns from the National Health Service Central Register, the Census of Employment (biennial from 1987), the decennial Census of Population and annual estimates of population change. Some of these sources are directly administered by the Department of Employment or its executive branches (including the Training Agency), others are the responsibility of

Table 10.2   Spatial aggregates available in NOMIS

Type of aggregation	Building blocks
Local Authority Districts	OPCS
County Council areas	District-based
Parliamentary constituencies	Ward-based
Statistical regions	District-based (office)
Travel-to-work areas	Office-based
CURDS functional regions	Office-based

Note: CURDS —   Centre for Urban and Regional Development Studies,
University of Newcastle upon Tyne.

Source: O'Brien *et al.*, 1988a

the Office of Population Censuses and Surveys or the Office of the
Registrar General for Scotland. Table 10.3 lists the current information
base maintained by NOMIS and the updating interval for each of the
major data sources.

Table 10.3   Data held on NOMIS as of July 1989

Data type	Source	Spatial unit	Updated
Unemployment	DE	DE ward	monthly
Employment	DE	DE ward	Triennially (Biennially from 1987)
Vacancies	DE	Job centre	monthly
Population	OPCS	OPCS ward	Decennially

Note: DE — Department of Employment
OPCS — Office of Population Censuses and Surveys

Source: O'Brien *et al.*, 1988a and NOMIS Newsletter, July 1989

NOMIS operates a continuous on-line computing service which may be
accessed interactively using communications networks such as British
Telecom's Packet Switch Stream (PSS) or the inter-university Joint
Academic Network (JANET). Dial-up facilities also allow access over
the telephone network from any part of the United Kingdom. This design

of centralised computing system with networked users was chosen so that NOMIS could offer the best possible combination of computing and communications facilities. A design based on distributed software would have required the development of machine-specific software, or lowest-common-denominator software; both options are significantly less efficient than the chosen option.

Table 10.4 Some members of the NOMIS user base

UK central government users

Department of Employment
Department of Trade and Industry
Welsh Office
Industry Department for Scotland
Home Office Research and Planning Department
Central Statistical Office
Cabinet Office
Department of Transport

Commercial users

Morgan Grenfell
Laurie Grimley
JR Eve
British Telecom
Business Strategies Ltd., London
Ove Arup Partnership, London
Hillier Parker Research
Weatheral, Green and Smith

Academic users

University of Newcastle upon Tyne
University of Durham
Birkbeck College, London
University of Oxford
Oxford Institute of Retail Management
University of Wales, Cardiff Business School
University of Edinburgh
University of Cambridge

Source: NOMIS Newsletter, July 1989

NOMIS offers three distinctive types of service tailored to the needs of different clients. First, to central and local government agencies, NOMIS provides data which can help with the formulation and delivery of policy.

Second, to market researchers and commercial consultancies, NOMIS provides a tool for geographical targeting and market profiling. Third, for academics, NOMIS offers extensive teaching and research possibilities at marginal cost. Each of these services imposes different requirements on the system. The following sections briefly describe each service.

Policy formulation and programme delivery

Just under half of the NOMIS user base work in central and local government, mostly in departments and agencies concerned with labour market analysis, employment and training initiatives and physical planning at a variety of geographical scales. (A list of some of these agencies is given in Table 10.4.) For the listed agencies, NOMIS offers a rapid and flexible means of accessing small area statistical data appropriate to their needs. In the case of the Training Agency, these needs are primarily concerned with documenting the demand for, and effectiveness of, government employment and training initiatives at highly localised levels (for example, wards or small aggregations of wards which may be used with policies aimed at inner-city regeneration).

Since 1988, the principal job of the Training Agency has been administering the Employment Training Scheme. This is a national initiative which is designed to ensure that all the long-term unemployed and school leavers not progressing to work or further education can receive some form of job training. The scheme involves identifying the priority groups for training (those aged 18-24 who have been unemployed for six months or more), developing a personal action plan for each individual based on his work aspirations and attempting to place the 'trainees' with approved Training Managers (often local businesses) who will deliver the work experience. Crucial to the success of this system is the ability of the locally-based Training Agents — organisations subcontracted by the Training Agency to administer Employment Training in a given area — to document the state of their local economies (Training Agency, 1988). NOMIS offers them an ideal tool for keeping abreast of the dynamics of their local economies.

Area targeting

The ability to access information for individual small geographical areas or their aggregates is particularly valuable to users whose operations involve some form of market analysis — that is, the selection of areas on the basis of some specified socio-economic or locational criteria (DOE, 1987). Such conditional selection is available within NOMIS.

All fifty or so data sets held on NOMIS may be accessed conditionally.

Simple conditional rules based on variables such as age-bands and sex, direction of migration, industrial and occupational categories, length of unemployment and type of unemployment, are available in most data sets. However, the combination of relationships which may be extracted depends on how the data sets are organised. Some incorporate a large number of structural relationships, allowing considerable scope for conditional searching — others are less complex and offer fewer opportunities. Information for the areas selected conditionally can be presented as tables or graphs, or stored externally for use in statistical analysis packages or spreadsheets. Figures 10.1 to 10.3 illustrate some of the graphical facilities available, based on three different conditional searches. Figure 10.1 displays the distribution of the elderly by electoral ward in East Sussex: the figures are taken from the 1981 Population Census. Figure 10.2 displays the pattern of in-migration to County Durham by young males aged 20 to 24 using data from the National Health Service Central Register. Figure 10.3 shows the ten worst counties in Great Britain for long-term unemployment as measured in June 1989. Conditional searches such as these are easily obtained and graphed using the mapping interface to GIMMS (Carruthers and Waugh, 1988) developed by the NOMIS team.

Academic usage

A third important client group served by NOMIS is academia. Two distinct uses of NOMIS are made by academics. First, many use the system for individual research and postgraduate teaching. Second, a small number of universities and polytechnics use NOMIS for undergraduate teaching. The former use is catered for by a simple extension of the service offered to government and commercial users. The latter requires the provision of a special bookable teaching account which can be used to obtain data for practical classes and seminars. The information holdings are widely used by geographers, economists, planners and sociologists, as sources of contemporary data for theory courses. However, NOMIS is also used in computer techniques courses as an illustration of an on-line networked computer system. The advantage of NOMIS is that it provides students with hands-on experience to large-scale data sets which they would not be able to get otherwise. Moreover, it allows them to experiment with plan making, selecting objective functions and testing alternatives. The ability to see the spatial consequences of their arbitrary and semi-arbitrary decision-making is an extremely important lesson in practical planning.

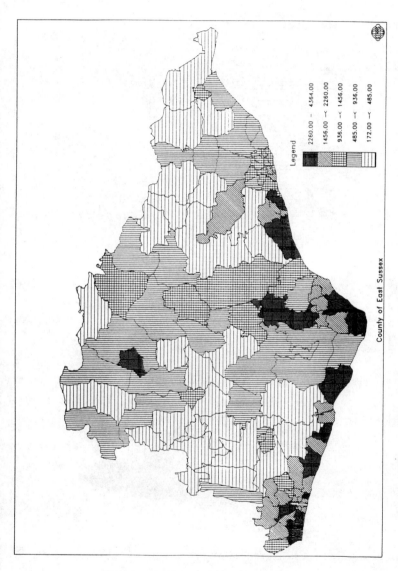

Figure 10.1 *Elderly by ward in East Sussex*

Source: OPCS Statistics (NOMIS)

Legend

2260.00 – 4,364.00

1456.00 –< 2260.00

936.00 –< 1456.00

485.00 –< 936.00

172.00 –< 485.00

County of East Sussex

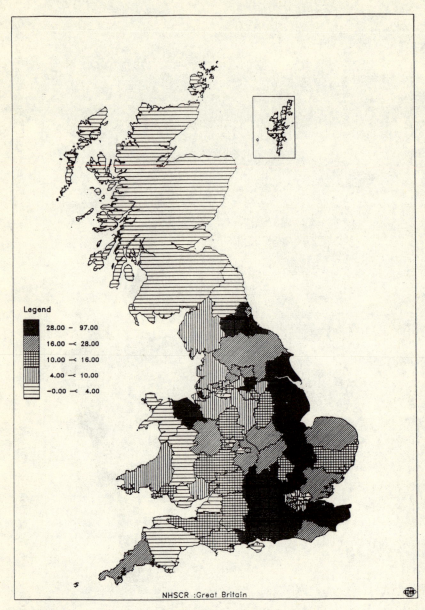

Figure 10.2 *In-migration to Durham March 1988-March 1989*

Source: OPCS Statistics (NOMIS)

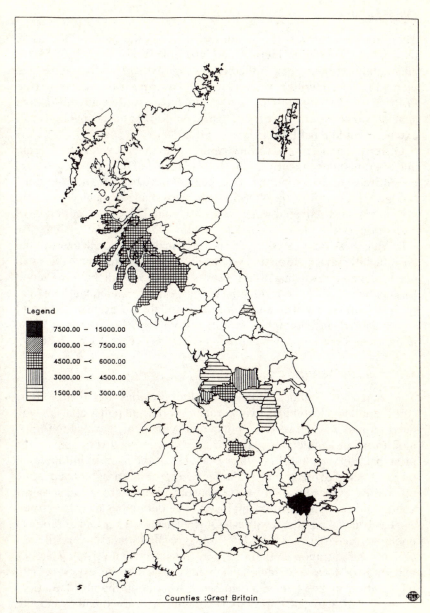

Legend

▓	7500.00 – 15000.00
▨	6000.00 –< 7500.00
▦	4500.00 –< 6000.00
▤	3000.00 –< 4500.00
▤	1500.00 –< 3000.00

Counties :Great Britain

Figure 10.3 *Ten worst counties for long-term unemployment in June 1989*

Source: D.B. Statistics (NOMIS)

Other facilities

The three client groups are very different. Academic and business users tend to be computer literate and are able to operate the system adequately without great difficulty. To assist these, a comprehensive manual has been written which covers all key aspects of the system (see O'Brien *et al.*, 1988a). This is updated regularly, but all substantive changes which occur before revisions are published are documented in an extensive on-line help and message system.

Government users tend to vary. Some are extremely computer literate and operate independently of the NOMIS team in Durham, while others are entirely new to computing. Two facilities have been provided to assist this latter group. First, the NOMIS team has provided a menu-based help system which is activated when NOMIS is run. This introduces NOMIS and outlines the steps needed to obtain data from the system. By following these steps carefully, the novice can be led through many of the basic facilities available on the system. The on-line menu is available to all users but can be switched off if it proves to be inconvenient. The second form of help available is residential training based either in Durham or at the user's office. Two training courses are available covering introductory topics and advanced graphics facilities. Both are supported by training manuals and self-help open learning materials (O'Brien *et al.*, 1988b).

Practical applications

The range of facilities outlined in the previous discussion offers considerable opportunities to integrate research and policy analysis with decision-making processes, policy development, evaluation and monitoring at a variety of spatial scales. For central government — the principal users of NOMIS — the system lies at the heart of initiatives to target unemployment and training policies and programmes more efficiently, especially where these concern priority groups such as the young and the long-term unemployed. The definitions of these groups can easily be manipulated within NOMIS because its age and duration of unemployment data are stored in a disaggregate format. A versatile mix of age and duration definitions may thus be created, allowing the dynamics of local and regional labour markets to be exposed. This information makes it possible to see how 'rule-of-thumb' definitions created from aggregate information work out in local economies and thus creates a way for central government to respond to local needs.

The same flexibility applies to local government use which frequently involves comparative analyses, both with the previous performance of the area and with other authorities, some of whom are perceived as competitors for resources and inward investment. In particular, NOMIS makes it possible for local authorities to assess the components of economic change affecting their areas and place these in context. Simple accounting formulae such as shift-share analysis and location quotients are available as standard options in many of the NOMIS modules. These allow patterns of change in local economies to be compared with benchmark conditions which may be defined at county, region, or national scales. Recorded changes may then be assessed in the light of, for example, national economic change for the economy as a whole, with the contribution of the area's industrial mix or peculiar locational advantages being set out separately. This information is vital if the correct mixture of initiatives and incentives is to be developed for local areas. Without NOMIS to provide country-wide data, such analyses would be extremely difficult, if not impossible, to perform at local levels.

However, NOMIS is not simply a tool for routine data analysis and information provision, though it performs these tasks admirably. By linking together a wide variety of data sets on an even wider variety of spatial scales, it offers the possibility of new insights and perspectives to aid the creation of knowledge about local labour markets and sub-regional economies. Champion and Green have made considerable use of the system to study what they term the 'booming towns' of England and Wales (Champion and Green, 1985, 1987, 1988). These are the areas which have prospered under the economic policies of the 1980s as measured by changes in skilled and professional employment, relatively low levels of unemployment and rapidly rising house prices. The particular contribution of NOMIS to this work was in providing information on employment, unemployment and migration in a form which made it very easy to integrate with data collected from other sources. More recent expressions of this work have been incorporated in published and forthcoming research looking at the changing space-economy of the country and have contributed to the debate and illustration in terms of real geographical areas, of factors such as counter-urbanisation and the north-south divide (Champion et al., 1987; Townsend and Champion (forthcoming), Champion and Townsend (forthcoming)).

NOMIS design

Key considerations

In order to be considered successful, NOMIS has to satisfy the needs and aspirations of a diverse user base. The following are a number of its more important design criteria: it has to maintain up-to-date information, it has to provide access to 'historical' information, it has to be easy to use, it has to be available during normal office hours, it has to be accessible over a communications network from all parts of the UK and it has to add value to the raw information.

These are difficult criteria to meet and some of them may be conflicting. For example, the need to add value suggests that NOMIS should provide analytical tools which can model the temporal and spatial processes operating at local, regional and national levels. These tools can be particularly sophisticated and are likely to be used by only the most experienced users. Second, the need to provide access to contemporary and historical data may conflict with the need to offer a fully on-line networked service. The former may involve extensive data searching, the latter efficient interactive processing. Third, the need to keep the system on-line during office hours may conflict with the need to keep the information system up-to-date. As the system has to be on-line and fully interactive, the three areas of routine maintenance, system development and public relations have to be finely balanced. These criteria can only be met by an effective system design.

Basic design

Figure 10.4 contains an outline of the basic structure of NOMIS. The system is organised into a housekeeping module (MAIN) and a series of analytical overlays which are activated by the SLINKS and DRIVER modules. The purpose of the former is to check access and use of NOMIS. The purpose of the latter is to gather together parts of the system which are interrelated into distinct sets — for example, routines to produce analyses of specific data sets. The overlays appropriate to the analysis specified by the user are read into memory as needed. This minimises the overall memory profile of the system (currently less than 1 megabyte).

Effective system design involves separating the various tasks expected of NOMIS into discrete units and identifying the critical path which links them together. All systems have a critical path which provides the baseline for judging their success. For NOMIS, data storage and the speed of data retrieval in the form appropriate for the user are the main considerations. The command language has been designed to provide all the necessary checks for validity at the housekeeping level. This means that pointless processing is avoided, as is the need to search for data which are not maintained in the system.

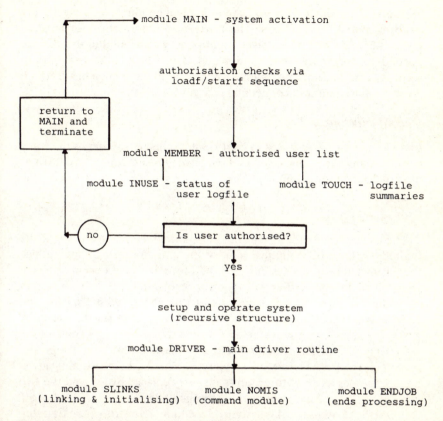

Figure 10.4 *Core modules in the NOMIS system*

Selecting data by area

Geographical searching lies at the heart of NOMIS. The system allows data to be selected for both individual building blocks or for spatial aggregations based on them. This means that each individual building block needs to be accessible by the system and be in a form which allows its contents to be easily aggregated.

The basic unit of storage used in NOMIS is a computer record which corresponds to a single geographical area. The length of each record depends on the data set being selected and varies depending on what information has to be stored. Each record is numbered and is linked to look-up tables which show how the individual building blocks aggregate to form spatial aggregations. The retrieval of information for most of these therefore merely involves identifying which building blocks are needed to create the aggregation and then reading the appropriate records. This data structure allows for very rapid retrieval, a factor which is essential to minimise connection charges for off-site users.

Record structure

Many of the data sets stored in NOMIS are sparse. This means that they contain many zero values, some of which may form sequences or strings of zero figures: for example

$$1\ 2\ 0\ 0\ 0\ 1\ 2\ 0\ 0\ 1\ 2\ 2\ 3$$

The main problem with storing zero strings is that they waste the limited resources of computer memory, increase storage costs and reduce the performance of the system. It is therefore desirable to remove them altogether and replace them with some form of dummy code. One system of coding is termed coded deltas (Visvalingham, 1976; Nelson and Blakemore, 1986; Blakemore, 1987) and involves replacing the strings by a negative number whose value depends on the number of zero values in the string. (Naturally, the dummy coding chosen must be appropriate; zero values may not be suitable for all types of data.) In the example given above, the sequence of three zeros may be replaced by -3 and the two zeros by -2; a reduction in overall storage of 25 per cent.

NOMIS makes use of this system of encoding zero strings as a first step in a two-stage data compaction process (Blakemore and Nelson, 1985). The second stage involves rewriting the line of data in assembler code. This is used because of the discrete nature of data storage in digital computers. In effect, the amount of storage needed to store a number depends on its size. However, there is a distinct ratchet effect between the numbers and the storage required which allows many large numbers to be stored in the same amount of space required for small numbers. The following conversions apply:

values up to 2047	12 bit storage
values between 2048 and 32767	16 bit storage
values between 32768 and 8388607	24 bit storage
values exceeding 8388608	32 bit storage

Thus, given that the largest value in the example above is 12, the whole line can be restored in assembler using 12 bit storage. If the largest value was 2048, the whole line would need to be stored in 16 bit storage, necessitating an increase in the amount of memory needed.

Table 10.5 shows the effect of this two-stage compaction process on some of the data sets held in NOMIS. In aggregate, compaction allows the raw data to be held in a tiny fraction of that needed otherwise. The effect on cost is considerable (£4.7 million reduced to £700k), though the storage charge is still high by any objective standard. Experiments in processing the compacted data show that to regenerate the raw data from the compact version adds relatively little to the overall storage costs and in no way offsets the cost savings generated from the exercise.

Table 10.5 Storage charges before and after compaction

Domain	Storage as % of original	Storage Full	charges Compact
Median duration of vacancies	1	1,785	28
Census of Employment	6	56	4
Age and duration data	8	681	59
Ward unemployment	38	45	17
Census of Population	58	38	23
Seasonally-adjusted unemployment	108	37	39

Note: Figures are in £k at 1989 unless specified otherwise

System appraisal

The ability to reproduce data in table or graphical form is one thing, it is quite another to analyse or interpret that information. There is an unfortunate tendency for users to think that tables and graphics are the product of research, whereas in fact they are often only the basis for research. Both are forms of data in their own right and both are highly problematic forms of data (Sayer, 1984; O'Brien, Blakemore and Townsend, 1988, O'Brien, 1990). An information system which merely produces tabulations and graphics is unlikely to be successful unless it can be integrated into an existing organisational decision-making structure which provides the context for interpretation (Batty, 1988). This section looks at some of the issues underlying an appraisal of NOMIS.

NOMIS as orgware

Nijkamp and de Jong (1987) have invented the term 'orgware' to refer to computer technology which has been designed to fit into existing organisational structures. Their purpose is to demonstrate that computer applications have to take place in the context of existing administrative and policy-making procedures and that, consequently, many applications are likely to fail to achieve their potential if their degree of fit is poor. In the past, attempts to incorporate computer systems in policy-making applications have often been ineffective because of the mismatch between the needs of the organisation and the facilities being offered by computers.

NOMIS is successful as orgware because its design has been worked out jointly between computer specialists and users of small area data in the Training Agency. The tasks given to the computer specialists were to reproduce many of the routine data gathering and report writing tasks that traditionally occupied Training Agency staff. With the system now providing an input to these tasks, more time has been released for staff to devote to interpreting the data. NOMIS has not attempted to replace these staff by providing some form of expert system which interprets the data automatically. Not only would such a system be out of place in the largely non-computerised areas of policy-making within the Training Agency, but it would be computationally inadvisable given the lack of detailed knowledge on the long-term performance of such systems (Watts, 1989).

NOMIS for data analysis

Data analysis, which is defined as the process of turning data into information, comprises both technical processing and social understanding. Both aspects are essential and neither one can be omitted. The essential technical aspects of data analysis have been described by Ehrenberg (1984) as comprising five key elements: a statement of the size of the sample or population; a measure of central tendency; a measure of dispersion; a description of the shape of the data; and a description of any irregular variability in the data. These elements may lead to the development of simple descriptive measures for single areas or variables or to measures of association between two or more variables.

Table 10.6 contains details of some of the measures which have been built into NOMIS. Virtually all of these are simple descriptive measures which allow sequences of data to be processed. As NOMIS is primarily a data retrieval system, the level of analytical facilities has been kept

Table 10.6   Measures of description and association available through
NOMIS

Description

Minimum and maximum of data
Rounding to specified integer
Truncation by absolute number and percentage
Ranking
Sorting
Average unemployment over a time period

Association

Location Quotient
Shift-share analysis
Likelihood of becoming or ceasing to be unemployed
Signed chi-square analysis
Median duration of vacancies analysis
Projected population estimates

---

modest. Simple descriptive measures have been included to rank and sort data by raw numbers or derived measures such as percentages and chi-square statistics. Truncation facilities allow subsets of areas such as the top and bottom 20 per cent to be selected. A number of more complicated descriptive measures have also been programmed into NOMIS. These include facilities to compute location quotients and to perform shift-share analyses which allow the economic performance of specific small areas to be compared with benchmarks such as the national economic performance. The most complex analyses are associated with the median duration of vacancies data sets which involve measures of temporal analysis which have been designed specifically for this data type.

Though NOMIS is primarily concerned with data retrieval and simple descriptive analysis, its data holdings may be the basis for complex space-time modelling. This poses a problem for NOMIS because the incorporation of new facilities increases the memory overheads needed to keep the system active. The route which has been followed to date is to program into NOMIS only those facilities which are likely to have general application or those which are needed by the Training Agency and to leave specialised developments to users. However, the development of the NOMIS spreadsheet has allowed for greater flexibility. This system, which is rather more like a MINITAB worksheet than a commercial spreadsheet, allows data to be extracted from different data holdings within NOMIS and combined in the same analysis. The ability to integrate these diverse holdings is very important and is a major factor in the ability of NOMIS to add value to the existing data. Furthermore, the spreadsheet allows data collected externally to be incorporated in any analysis with NOMIS data. A typical example of this might be to link NOMIS employment data with a recent local area demographic survey. This overcomes the inherent problem that the small area population data held on NOMIS are now nearly ten years old.

NOMIS input and output

The main objective of modern approaches to computing is to link together computer systems which have been designed for specific tasks. This differs from earlier attempts to develop all-embracing general purpose systems which usually failed to meet the full needs of any of their users. NOMIS, though large in its conception, is actually a relatively small computing system. It contains only a minute amount of official government data and an even smaller amount of the small-area geographically-referenced data which is available in the United Kingdom. A comprehensive analysis of small areas will thus involve obtaining data from NOMIS and from other sources. If NOMIS is to be a successful component in a networked information system, users have to be able to input commands and output data without needing to invest too heavily in system specifics.

The input system incorporates three types of command: network commands to link the user with the Durham University computer which is part of the Northumbrian Universities Network (NUNET), operating system commands to manipulate files and messages before running NOMIS and NOMIS-specific commands which are needed to access data and perform analyses. This threefold division is not complex but can be confusing to novice users, especially as the responses which are displayed on the user's terminal may not always make it obvious which part of the system produced them. Network connection is relatively easy because NUNET is registered with both PSS and JANET and it provides several telephone numbers for dial-up access. Distinguishing between operating system and NOMIS commands is a matter of experience, though the training materials and the manuals devote considerable attention to these issues.

Input is an expensive operation, but one which can be controlled. Costs may be minimised by storing all operating system and NOMIS commands in a file and routing all output to a second file. This allows users to limit their connection to NOMIS to a single file handling operation. A development of this is the ability to create the files locally and then send them to the Durham site for processing. This allows costs to be reduced still further as the process of creating workable NOMIS command sequences could be handled locally using a command parser which does not even attract network connection charges. Both processes seek to minimise network time.

As for output, apart from government users who regularly require hard copy output, the trend is towards networked output, with data and graphics being copied across PSS and JANET for processing locally. Data transference is particularly easy, as NOMIS provides facilities to copy both transcriptions of the computing session and statistical information directly to the user's own computer. If graphics are needed, NOMIS can be instructed to structure the output in a form which can be read directly by GIMMS or pcMapics. The only difficulty with graphics is that the user must have access to digitised boundaries. The boundaries stored by NOMIS for processing of graphics in Durham cannot be copied down over a network.

Other concerns

Any successful computer system has to be able to respond to a wide range of users doing a wide range of tasks. These need to be accommodated by the command language and the entry and exit facilities. NOMIS succeeds to the extent that it has been designed within the context of a computing network and therefore facilities which are really nothing to do with NOMIS (for example, the use of modem equipment and university screen editors) have been incorporated in its documentation. A successful system has to try to meet the needs of its user base, though being mindful of the needs of its primary client. NOMIS has attempted to do this by setting up a system to process suggestions for new facilities or complaints about existing facilities. This has led to modifications which have helped the analytical tasks of many users. By providing this and a regular information exchange through newsletters and working papers, the system attempts to match its facilities with its user base, as well as expand its user base by marketing its facilities.

Fundamental design issues

NOMIS is generally considered to be successful because it meets the needs of its primary client and makes the effort to find out the needs of its other clients. It also does not claim to be more than it is. However, for information systems which are aimed at different markets, or at markets which are ill defined (such as the market for special needs information), these considerations, though necessary, will not be sufficient. This section expands on some of the comments raised previously by looking at some of the more fundamental design issues involved in creating small area information systems.

One of the key measures of success of an information system is its ability to produce sensible information which can lead to sensible policy-making. Unfortunately, the level of knowledge required for users of geographically-based systems is rather high. While geography is easy to accommodate in a computer system, inferences based on geography can be very complex. The following are some of the key difficulties which will affect the interpretation of any form of tabular or graphical output from a GIS: spatial units are not fixed entities but can be modified; there is no basic spatial unit which is universally accepted; measures of association developed on spatial units will reflect those units; classifications based on spatial units may be particularly responsive to minority characteristics; and patterns of associations may not have a causal basis. Three of these aspects are worth considering in more detail: these are spatial unit problems, survey analysis problems and measurement problems.

Spatial unit problems

The effect of variable spatial units is best demonstrated by example. Figures 10.5 and 10.6 show the proportions of elderly households to all households in 1981. The values in Figure 10.5 suggest that the main elderly areas are the South Coast, excluding Hampshire, most of mid and North Wales, South-East Scotland and the Western Isles. Figure 10.6, in comparison, suggests a rather different picture, with concentrations of the elderly along much of the coast, as well as throughout mid-Scotland and the North-east. The reason for these differences in representation is that the spatial units used in the two figures have changed the way the information is described. In Figure 10.5, 66 counties have been used to categorise the data, whereas over 450 local government districts are used in Figure 10.6. The proportions produced from the same raw data therefore reflect the choice of spatial unit as well as the relationship between the two variables. The key analytical problem here is that it is difficult to determine how much of each proportion is due to the effect of the variables or of the spatial units. As the choice of counties or districts may be arbitrary, it is not clear which set of findings should be used to represent the patterns in the raw data.

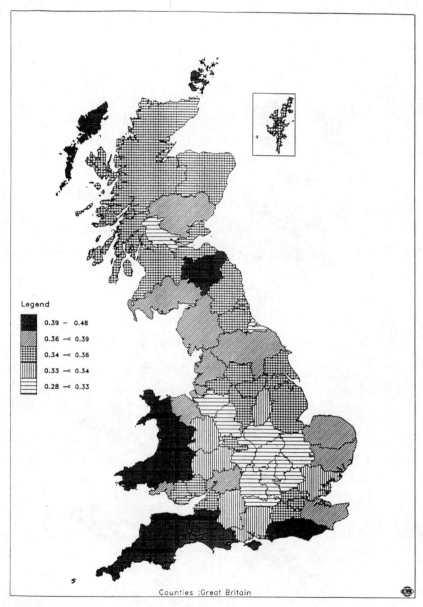

Figure 10.5 *Proportion of elderly households to all households by county (1981)*

Source: OPCS Statistics (NOMIS)

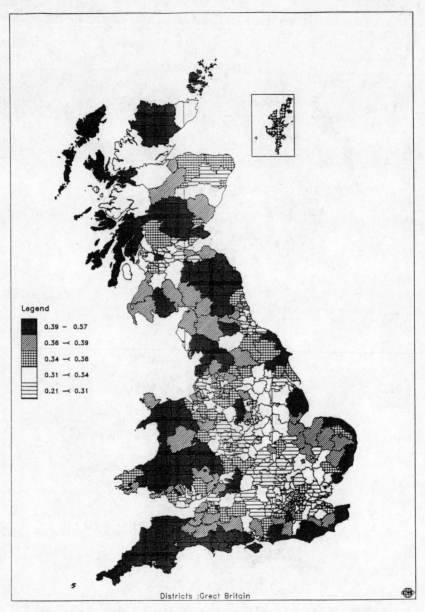

Figure 10.6 *Proportion of elderly households to all households by district (1981)*

Source: OPCS Statistics (NOMIS)

Two related issues arise from this problem of modifiability. First, users need to decide which spatial units are to be used. This choice may be dictated by administrative procedures (for example, by an authority always using wards) or it may be *ad hoc*. In either situation, the users need to determine whether the interpretation of the information at the chosen spatial level is adequate for the policy question under consideration. In many cases it will not be simply because units will be described by minority characteristics. For example in Figure 10.1, which shows the relative concentration of the elderly by ward, there are very few wards in which the elderly form the majority of the population. To talk therefore of the coastal belt as being a 'Costa Geriatrica' is to misrepresent the spatial information.

Second, having chosen the units to be used, the analyst needs to determine their shape and number. Both aspects affect how the variability in the data is described. For example, should ten or twenty units be used? If the research suggests ten units, should these be equally sized by area or population? Or should they be the same shape? These questions can be answered formally and systematically by applying procedures of numerical taxonomy and pattern matching to the collected data. However, the intellectual and political costs of adding these tools to an administrative system may be too great for that system to bear.

These problems are recognised by most geographers and planners who regularly analyse spatial data but there are many who use spatial information without recognising the issues. In the absence of an agreed basic spatial unit or methods of spatial aggregation and division, the chances of information systems producing outrageous inferences must be considered high. A detailed discussion of these issues and some of the solutions on offer is provided by Openshaw, writing in Appendix 7 of the Chorley Report (DOE, 1987).

Survey analysis problems

Small area information systems such as NOMIS tend to be built on existing secondary data sources rather than on data collected by the users themselves. This means that they are susceptible to all the flaws and idiosyncrasies of such sources (Moser and Kalton, 1971; Short, 1980; Marsh, 1989). Two problems are particularly important: coverage and non-response. The former affects all types of secondary statistical source, even those which claim to be enumerations of the whole population. The latter affects the interpretations which can be placed on information gathered using sample surveys. Many government statistical collections now involve sampling rather than enumeration and so suffer from sampling deficiencies.

Sampling is a very effective way of obtaining information on a population without actually having to study it. The rules of statistical sampling are well known (see Moser and Kalton, 1971) and if appropriately applied, ensure that inferences are valid. However, sampling theory makes several assumptions about the characteristics of the population which may not be realistic in all situations, for example, the idea that the study variable is irregularly distributed and large in number. Many geographical distributions violate this idea simply because the social and economic nature of our society tends to concentrate people who share characteristics in given localities: for example, the professional classes tend to be concentrated in modern estates of owner-occupied housing designed for the two-car family. More practically, sampling assumes that contemporary, unbiased lists exist which can be used for sampling. These should be up to date and be a complete coverage of the area and study variable. These are unrealistic assumptions. Most lists are out of date and have entry rules which exclude certain categories of people (the electoral roll excludes all non-British people except the Irish). Even the complete enumeration of the Population Census does not claim to be 100 per cent accurate, nor can it be up to date when it is published. It is most deficient in recording information on inner-city problems, multi-occupation, ethnic matters and demographic data on the very young. These problems are also replicated in locally derived survey material which may be analysed in conjunction with NOMIS data (see Worrall, 1989, p.28).

This suggests that all lists which might be used as the information base of a small area information system may be inadequate in some way. The problem is in what way? One of the factors which leads to incomplete coverages is the presence of non-respondents in the survey. These are people who have been selected from the sampling list for inclusion in the study but who fail to participate. There are many reasons for non-response: for example, the respondent does not wish to take part or cannot be found by the surveyors. Regardless of the reason, non-response is a major problem because it affects the confidence which can be attached to inferences based on the survey statistics particularly if non-repsonse is not random and the sample is, to some (unknown) degree, self-selecting. If the sample is peculiar in some way, there is a real chance that the data will also be peculiar. Similarly, a survey with large numbers of non-responders may provide information which is essentially meaningless, especially if the source of the non-response cannot be determined. If meaningless data forms the basis for an information system, no amount of clever techniques will make the system successful.

GIS-generated 'pretty pictures' and banks of technology are no substitute for carefully designed data collection procedures and critical thought.

Measurement problems

All information systems must face the question of what their data means. This is not merely a comment about spatial representation — it concerns the social content of the data and, in particular, what the data mean in the context of peoples' lives. There is a tendency to assume that the information being processed by a computerised information system must be valid, especially if it has been collected by official bodies such as local authorities and central government. However, validity may range from truthfulness to consistency. The former refers to statements which correspond to real features of the world. The latter refers to statements which correspond to administrative processes measuring those features. The two may coincide, but there is no reason why they should.

The measurement of a social and economic variable is itself a social process because it reflects perceptions and assumptions on the part of the user which may not be fully sustainable — basically, classification systems in the social sciences are not value-free. Terms used to record the data at source may owe more to administrative convenience than to research needs. This means that they can misrepresent the data rather than illuminate it. Many socio-economic terms used in information systems are examples of 'chaotic conceptions' (Sayer, 1984). These are largely catch-all classifications and measurements used to place phenomena which are only marginally related. Such classifications compartmentalise the data with little regard for their structure or form.

Chaotic conceptions arise very frequently when the object of study lies across departmental boundaries and is thus perceived differently by the various agents involved. A good example of a chaotic conception which hinders the creation of small area information systems for people with special needs is the term 'disability'. Blaxter (1976) suggests that two partially incompatible definitions are paramount in British studies of the disabled. First, there is the approach which sees the disabled as 'innocent victims' requiring assistance, either from the state or charity. Such a view appears to have predominated in the social welfare studies in the nineteenth and early twentieth centuries when many of the associations for the blind and deaf were established. Second, there is the view that the disabled person is one who is 'substantially handicapped in obtaining or keeping employment, or in undertaking work on his (*sic*) own account, of a kind which apart from his infirmity, disease or deformity, would be suited to his age, experience and qualifications' (Disabled Persons (Employment) Acts 1944-58).

The former characterises the disabled as persons whose bodies do not work correctly. The latter characterises them as persons who cannot work. Friedson (1965) goes further than Blaxter by noting that disability is frequently seen as a form of social deviance. In this context, a person is disabled because he or she is seen to be a member of a stigmatised or deviant category; a category whose very existence reflects the commonly-held, but rarely stated, perceptions of society about what is or is not 'normal'.

The most recent official studies on the incidence of disability in Britain (OPCS, 1988a,b; 1989) make use of a medical definition. This is based on the classification recognised by the World Health Organisation in its International Classification of Impairments, Disabilities and Handicaps (ICIDH). Impairment, disability and handicap are distinguished as three possible outcomes of disease which correspond to:

1.   impairment: a situation when parts or systems of the body do not work properly;

2.   disability: a situation where people cannot perform tasks to normal expectation; and

3.   handicap: a situation where people cannot function in relationship to their environment or socially.

In effect, the classification distinguishes between conditions which are essentially medical (impairments) and those which are socio-economic (handicaps). Disabilities fall somewhere in between in that they cover a range of medical conditions which may or may not be handicapping depending on the ability of the individual to overcome socio-economic and environmental obstacles.

The problem of these diverse definitions is that they are not easily reconciled. The ICIDH definition, for example, could be used to include women with children among the handicapped, because their ability to function in relationship to their environment or society is restricted by social conventions and inadequate crèche arrangements. While these people may not be disabled in a medical sense, their functionality as determined by society is similarly restricted. It therefore may be reasonable to include them in a special needs/disability information system. This would be quite a different system from one based on medical or social welfare criteria.

While this discussion of classification issues as they relate to the disabled may appear esoteric, it is a problem which relates to most socio-economic and socio-demographic phenomena and particularly to constructs such as unemployment, employment and economically active. The application of GIS in practical policy-making environments is particularly susceptible to this type of problem and it a problem about which all policy analysts and decision-makers should be aware — few are.

Conclusions

The purpose of this chapter has been to discuss the problems and prospects of developing successful small area geographic information systems. NOMIS has been used to illustrate some of the factors which need to be considered. Most important of these is the need to know the client base and the sorts of operations clients will want to perform. NOMIS is successful because it has identified a market for itself within one small area of government activity and has used this as a springboard to develop other products and applications. NOMIS has not tried to replace the existing administrative systems but has added value to them by automating routine tasks: expensive human resources can now be more cost-effectively applied to the consumption and use of information rather than its production.

The success of NOMIS should not be taken to mean that all small area geographic information systems will be successful. Though there is no shortage of interest in their development, with many local authorities, government departments and businesses investigating their potential, off-the-peg software has a poor record of success and potential users have every right to be wary. Say's Law — supply creates its own demand — may be the watchword of the booming information systems industry, but most potential users are more aware of the well-known GIGO syndrome.

There are real problems with developing any form of integrated information system, let alone one which can be meaningful for small areas. First, an information system requires data. In many cases data will exist but in others nothing usable will be available. This is very likely to be the case for small areas, suggesting that extensive surveys or new methods of data collection or data synthesis will be needed (see Birkin *et al.*, elsewhere in this volume). This has cost implications and it may also require new types of local government structures. Second, the needs of small area systems may conflict with those of a national system. This may be because the definitions and assumptions underlying the former are fundamentally different from the latter. If so, whose views should prevail? Third, information systems at all levels provide business and government with unprecedented access to information (however accurate or truthful) on identifiable small areas. To what extent do the users of these systems recognise the inherent inferential problems posed by spurious correlation and rudimentary pattern matching? Correlation is not causation; finding the existence of a pattern is not in itself meaningful. The identification of artefactual patterns in data may also deflect the unwary analyst away from a concern about the social processes which created the data, an understanding of which are far more important inputs to policy development. These are factors long recognised by geographers, but are likely to be among the first casualties when geographical research techniques become office techniques. Finally, access to information has political implications. A system which is favoured solely for its technical achievements may not provide users with an effective tool for studying society. If information systems are to be used to target services and guide government, they need to be research tools as well as administrative tools. Systems which merely generate quickly large quantities of inaccurate information, misclassifications and spurious socio-economic conceptions, may serve to disenfranchise the public they were originally designed to serve. This is crucial. The prospects for geographic information systems developments are good, but the problems are enormous. They cannot be ignored.

Bibliography

Batty, M. (1988), 'Informative planning: the intelligent use of information systems in the policy-making process', *Technical Reports in Geo-Information Systems, Computing and Cartography,* 10, Wales and South West Regional Research Laboratory, PO Box 906, Cardiff, CF1 3YN.

Blakemore, M.J. (1987), 'Data compaction on the UK National Online Manpower Information System', Paper presented to the SORSA Conference, University of Durham, UK.

Blakemore, M.J. and Nelson, R. (1985), 'Data compaction in NOMIS: a geographic information system for the management of employment, unemployment and population data', *University Computing,* 7, pp.144-47.

Blaxter, M. (1976), *The meaning of disability,* Heinemann, London.

Carruthers, A.W. and Waugh, T.C. (1988), *GIMMS reference manual edition 5,* GIMMS Ltd., 30 Keir St., Edinburgh, EH3 9EU.

Champion, A.G. and Green, A.E. (1985), *In search of Britain's booming towns,* Discussion Paper 72, Centre for Urban and Regional Development Studies, University of Newcastle upon Tyne.

Champion, A.G. and Green, A.E. (1987), 'The booming towns of Britain: the geography of economic performance in the 1980s', *Geography,* 72, pp.97-108.

Champion, A.G. and Green, A.E. (1988), *Local prosperity and the north-south divide: a report on winners and losers in 1980s Britain* University of Newcastle, Newcastle.

Champion, A.G., Green, A.E., Owen, D.W. Ellin, D.J. and Coombes, M.G. (1987), *Changing places: Britain's demographic, economic and social complexion,* Edward Arnold, London.

Champion, A.G. and Townsend. A.R. (forthcoming), *Contemporary Britain: a geographical perspective,* Edward Arnold, London.

Department of Environment (DOE) (1987), *Handling geographic information,* HMSO, London.

Ehrenberg, A.S.C. (1984), *A primer in data reduction,* Wiley, London.

Friedson, E. (1965), 'Disability as social deviance', in M.B. Sussman (ed), *Sociology of disability and rehabilitation,* American Sociological Association, Washington DC.

Leith, P. (1989), 'The alchemy of dross', *The Times Higher Educational Supplement,* 2 June.

Marsh, C. (1989), *Exploring data,* Polity Press, London.

Moser, C. and Kalton, G. (1971), *Survey methods in social investigation,* Gower, Aldershot.

Nelson, R. and Blakemore, M.J. (1986), 'NOMIS — a national geographic information system for the management and mapping of employment, unemployment and population data', *Technical Papers,* ACSM-ASPRS Annual Convention, Vol 1, pp.20-9.

Nijkamp, P. and de Jong, W. (1987), 'Training needs in information systems for local and regional development and planning in developing countries', *Regional Development Dialogue,* 8, pp.72-119.

OPCS (1988a), 'The prevalence of disability among adults', OPCS surveys of disability in Great Britain Report 1, HMSO, London.

OPCS (1988b), 'The financial circumstances of disabled adults living in private households', OPCS surveys of disability in Great Britain Report 2, HMSO, London.

OPCS (1989), 'The prevalence of disability among children', OPCS surveys of disability in Great Britain Report 3, HMSO, London.

O'Brien, L.G. (1990), *Generalised linear modelling in geography,* Routledge, London.

O'Brien, L.G., Blakemore, M.J. and Townsend, A.R. (1988), 'It depends how you look at it', *Geographical Magazine,* Analysis supplement, 12, pp.5-7.

O'Brien, L.G., Nelson, R., Dodds, P. and Blakemore, M.J. (1988a), *NOMIS reference manuals* (4 volumes), NOMIS, University of Durham.

O'Brien, L.G., Nelson, R., Dodds, P. and Blakemore, M.J. (1988b), *Introductory users guide,* NOMIS, University of Durham.

Rhind, D. and Mounsey, H. (1988), 'GIS/LIS in Britain in 1988' in P. Shand and R. Moore (eds), *The Association for Geographic Information Yearbook 1989,* Taylor and Francis, London.

Sayer, A. (1984), *Method in social science: a realist approach,* Hutchinson, London.

Short, J.R. (1980), *Urban data sources,* Butterworths, London.

Townsend, A.R. Blakemore, M.J. and Nelson, R. (1987), 'The NOMIS database: availability and uses for geographers', *Area,* 19, pp.43-50.

Townsend, A.R., Blakemore, M.J., Nelson, R. and Dodds, P. (1986), 'The National Online Manpower Information System (NOMIS)', *Employment Gazette,* February, pp.60-4.

Townsend, A.R. and Champion, A.G. (forthcoming), 'The urban-rural shift', in R. Martin and P. Townroe (eds), *Regional development in the British Isles in the 1990s,* Kingsley, London.

Training Agency (1988), *Achieving quality: approved Training Agent's criteria information pack,* Training Agency, Sheffield.

Visvalingham, M. (1976), 'Indexing with coded deltas: a data compaction technique', *Software — Practice and Experience,* 6, pp.397-403.

Watts, S. (1989), 'Expert systems can seriously damage your health', *New Scientist,* 1665, p.35 (20 May).

Worrall, L. (1989), 'Urban demographic information systems', in P. Congdon and P.W.J. Batey (eds), *Advances in regional demography: information, forecasts and models,* Belhaven, London.

Index